6/2
Phil

POPPER, OBJECTIVITY AND THE GROWTH OF KNOWLEDGE

Also available from Continuum:

Descartes and the Metaphysics of Human Nature, by Justin Skirry
Kierkegaard's Analysis of Radical Evil, by David A. Roberts
Nietzsche and the Greeks, by Dale Wilkerson
Rousseau's Theory of Freedom, by Matthew Simpson
Leibniz Reinterpreted: The Harmony of Things, by Lloyd Strickland
Popper's Theory of Science, by Carlos Garcia
Dialectic of Romanticism, by David Roberts and Peter Murphy
Tolerance and the Ethical Life, by Andrew Fiala
Aquinas and the Ship of Theseus, by Christopher M. Brown
Demands of Taste in Kant's Aesthetics, by Brent Kalar
Philosphy of Miracles, by David Corner
Platonism, Music and the Listener's Share, by Christopher Norris
St Augustine and the Theory of Just War, by John M. Mattox
St Augustine of Hippo, by R. W. Dyson
Wittgenstein's Religious Point of View, by Tim Labron

POPPER, OBJECTIVITY AND THE GROWTH OF KNOWLEDGE

JOHN H. SCESKI

Continuum International Publishing Group
The Tower Building, 11 York Road, London SE1 7NX
80 Maiden Lane, Suite 704, New York, NY 10038

© John H. Sceski 2007

All rights reserved. No part of this publication may be reproduced or transmitted in any form or by any means, electronic or mechanical, including photocopying, recording, or any information storage or retrieval system, without prior permission in writing from the publishers.

British Library Cataloguing-in-Publication Data
A catalogue record for this book is available from the British Library.

ISBN: HB: 0-8264-8904-4
         9780826489043

Library of Congress Cataloging-in-Publication Data

Sceski, John H.
   Popper, objectivity and the growth of knowledge / by John H. Sceski.
      p. cm. – (Continuum studies in British philosophy)
   Includes bibliographical references and index.
   ISBN-13: 978-0-8264-8904-3
   ISBN-10: 0-8264-8904-4
1. Popper, Karl Raimund, Sir, 1902-1994. 2. Objectivity.
3. Science – Methodology. 4. Cosmology. I. Title II. Series.

B1649.P64S34 2007
192 – dc 22
                                                    2006028507

Typeset by Aarontype Limited, Easton, Bristol
Printed and bound in Great Britain by Biddles Ltd, King's Lynn, Norfolk

Dedicated with love to my wife Renee and to the blessings that keep us whole: Blaise, Collin, Magdalen, Dominic, Clare, and Eileen

Contents

Preface	x
Acknowledgements	xiii

1. Introduction: Overview of the Argument ... 1
   1. Problems of Philosophical Problem Solving: An Apologetic Beginning ... 1
   2. Overview of Popper's Solution to the Problem of Objectivity ... 14
   3. Comments on the Problem Situation ... 18
      - 3.1. The Epistemological Problem ... 19
      - 3.2. The Metaphysical Problem ... 21
      - 3.3. The Linguistic Problem ... 23
      - 3.4. The Political Problem ... 26
      - 3.5. The Ethical Problem ... 29

2. Scientific Method and Objectivity ... 33
   - Introduction ... 33
   1. Popper's Solution to the Problem of Demarcation ... 35
      - 1.1. The General Logical Context ... 35
      - 1.2. Empirical Refutability ... 37
         - a. Strictures on Scientific Testing ... 40
            - a.1. Intersubjectivity and Repeatability ... 40
            - a.2. Theory-ladenness and Conceiveability ... 41
      - 1.3. Consideration of Criticisms ... 47
         - a. Asymmetry ... 47
         - b. Empirical Basis ... 49
         - c. Repeatability ... 54
         - d. Kripke and Intersubjectivity ... 55
   2. The Problem of Induction ... 60
      - 2.1. The Solution: Conjectural Knowledge ... 60
      - 2.2. The Four Interrelated Problems of Induction ... 62
         - a. The Logical Problem ... 62
         - b. The Epistemological Problem ... 63

		c.	The Methodological Problem	66
		d.	The Metaphysical Problem: The Unifying Aspect of Popper's Theory of Objectivity	67
		e.	Popper's Solution to the Metaphysical Problem	69
	3.	Corollaries: Corroboration, Truth, and Verisimilitude		76
		3.1. Corroboration		76
		3.2. Truth		78
		3.3. Verisimilitude		81

3. Cosmology and Propensities — 90
   Introduction — 90
   1. Popper's Account of Propensities — 90
      1.1. Physical Propensities, Probability, and Tests — 93
         a. Popper's Early Treatment of Probability — 93
         b. Von Mises' Frequency Interpretation — 95
         c. Popper's Frequency Theory of Probability — 96
         d. The Propensity Theory of Probability — 97
   2. Evolutionary Epistemology — 103
      2.1. Neo-Darwinian Evolution: A Standard Account — 104
      2.2. The Logic of Evolution and Critical Rationalism — 105
         a. No Guarantee — 105
         b. Growth by Criticism — 105
         c. Neo-Darwinism and Criticizability — 105
         d. Historicity — 106
         e. Organisms are Problem Solvers — 107
         f. Genetic Basis — 107
         g. Situational Logic — 107
         h. Background Knowledge — 108
   3. The Fundamental Difference: Towards an Evolutionary Ontology — 109
      3.1. An Evolutionary Ontology — 112
         a. World Three and Its Interactions — 114
         b. The Objectivity of World Three — 115
         c. Objective Propensities Concurrent with Language — 117
         d. Propensities and Products — 118

4. An Objective Social Order: Politics and Ethics — 127
   Introduction — 127
   1. Evolution and the Myth of the Framework — 128
   2. Social Science Methodology — 135

3.	Historicism	137
4.	Ethics	139

Bibliography   149
Index   156

Preface

As far as humans can tell, objectivity is a problem for them alone. Like humans other organisms struggle to secure their place in the world; nevertheless, the intriguing questions of whether they are deceived by their coping mechanisms, the unique status of their subjective contributions in opposition to contributions from the environment, and how to situate themselves in light of such a distinction, never seems to engage the attention of other living things that share this planet with us. As well, it makes little sense to assert that God and the angels struggle with this problem. The former knows too much and concerning the latter too little is known. Moreover, God and non-human organisms share a marked similarity: in each case their knowing and the object of knowledge are one. God knows all things in knowing himself, and non-human organisms view their knowledge and the world as one until they are selected out, and then it is too late for them to know anything. Thus, our best guess is that the problem of objectivity, like many other problems, is distinctively human. To say this may not be very informative except that it highlights an important contrast: unlike God who does not make mistakes and unlike non-human organisms who make mistakes but do not recognize them as mistakes, humans both make mistakes about the objective status of things and are often painfully aware of their errors as errors. Thus the problem of objectivity arises for us as the occasion of error forces us to reflect upon the gap between our subjective commitments and a world recalcitrant to our 'best laid schemes'.

Accordingly, the problem of objectivity has ontological and epistemological dimensions. Concerning the former, a central question is what is it about the reality of things that entitles us to designate cabbages and computers and interior states such as beliefs and desires as both real, yet different? This is a question about the status of things as real and the distinctions that can be made between them; in philosophical jargon it is a question about the objective ontological status of various states of affairs. The epistemological dimension of the problem has been a special focus of philosophy since the Renaissance and distinctively characteristic of Enlightenment thought.

The epistemological issue concerns the defining features, scope, and veracity of human knowing in reference to our judgments about the world. Centrally, the epistemological problem of objectivity asks what the constraints on our knowledge claims are, and if there are any how do we know there are, and are any of these constraints mind-independent? These issues arise in the light of our recognition of human fallibility. As indicated above, we are often mistaken concerning what we claim to know and the responsibility is ours alone. Thus a major issue for us is how can we control error and what role does a mind-independent reality play in that control? The notion guiding Renaissance and Enlightenment philosophical endeavors is that to answer the epistemological problem of error control can help us to attain a favorable perspective on the status of things. However, contemporary philosophers of all persuasions, if not disenchanted with Enlightenment aims are disappointed with the epistemological and metaphysical arguments for realizing them. Moreover, the linguistic emphases of contemporary philosophers provide little help in addressing the problem of objectivity if only because they have rejected the possibility of objective problems.

Among Western philosophers, Karl Popper provides a singular contribution to the problem of error control and thereby his thought is uniquely situated to address both dimensions of the problem of objectivity. The aim of this work is a critical exposition of Karl Popper's philosophy placing special emphasis on his treatment of objectivity. To this end, I examine his general account of rationality and philosophical methodology (Chapter 1), his treatment of particular problems in the philosophy of science and cosmology (Chapters 2 and 3), and his arguments for an objective social order where political and sociological questions are addressed (Chapter 4). Finally, at the conclusion of the monograph I extend his treatment of objectivity to morals and provide a sketch of a genuine Popperian ethics. I assert that Popper's philosophy provides the best framework to answer all questions concerning objectivity: epistemological, metaphysical, political, linguistic, and ethical.

In the name of error control I have of course consulted the works of other thinkers, especially those who address Popper's contributions to philosophy. Among these David Miller deserves special acknowledgement, and I would add respect. Miller's writings merit serious attention not only as an exposition of Popper's thought but also for their resourceful development and extension of Popper's philosophy of critical rationalism. Miller deserves much credit for the distinctions between classifying statements as true and believing them to be true, and methodological skepticism and epistemological skepticism; additionally, he has gone farther than Popper in exploring the nature of deductive inference and assessing its contribution to rational inquiry; lastly, his technical expositions of key issues for critical rationalism (verisimilitude, inductive probability) have significantly advanced the debate in their respective domains.

However, in addition to engaging other philosophers who directly address Popper's thought, the text also brings Popper's philosophy into dialogue with key thinkers in the contemporary analytic and pragmatist traditions, namely Saul Kripke and Donald Davidson. Popper scholarship has sometimes been likened by its practitioners to a family squabble; nevertheless, I count myself among those thinkers who believe that Popper's philosophy has a home in the broader philosophical debate. It is part of the aim of this text to argue that Popper's philosophy merits a more extensive place in contemporary philosophical discourse than it has generally received.

Concerning the text, I inform the reader that unless otherwise noted the use of italics belongs to the author(s) cited in the references.

Acknowledgements

I would like to thank Richard J. Blackwell and John A. Doody for their kind assistance in the completion of this project. The author of this monograph reserves all errors. I would also like to thank Todd Wilmot of Ryan Memorial Library for assistance with research materials and Sarah Douglas, Series Editor at Continuum Press, for her kind patience. Moreover, I am grateful to Sir Karl Popper's literary executor, Mrs Melitta Mew, and to Routledge Press for permission to quote from Popper's works. Most importantly I would like to thank my wife Renee and our children for their loving support throughout this project. As indicated in the dedication, this book is for them.

Chapter 1

Introduction: Overview of the Argument

1. Problems of Philosophical Problem Solving: An Apologetic Beginning

Philosophers take it to be their special province to ask questions. However, for some philosophers less is more: philosophy at its best is the attempt to undo philosophy by delivering its subject matter into the hands of other disciplines[1] or it involves the recognition that no genuinely philosophical questions can be asked.[2] When the later-Wittgenstein states, 'the real discovery is the one that makes me capable of stopping doing philosophy when I want to. – The one that gives philosophy peace, so that it is no longer tormented by questions which bring itself into question',[3] he identifies an aim that he simultaneously acknowledges cannot be satisfied. The aim Wittgenstein seeks is complete clarity such that philosophical problems '*completely* disappear';[4] however, if the 'real discovery' is to realize the requisite clarity that justifies the removal of philosophical problems, then on Wittgenstein's own reckoning such clarity is trivially omnipresent and at the same time insufficient to constrain philosophical questioning. Concerning the philosophical landscape, 'since everything lies open to view there is nothing to explain'[5]: 'nothing is hidden'[6] because the problems of philosophy are problems of language and since language is our own creation it is therefore amenable to easy survey. Yet despite the conspicuous nature of the philosophical panorama the real discovery that will eliminate philosophical questioning is not to be had.

According to Wittgenstein, because the perspicuity of the philosophical landscape is not at issue; the problem is how philosophers view it. The traditional approach is to view language as denoting essences, or as Wittgenstein argues in the *Tractatus*, language pictures facts. Concomitant with this essentialist approach to language is an essentialist approach to methodology. If there is only one thing that language does, it captures the essence of the things we speak of and thereby fixes their meanings, then philosophical problem solving only requires one method to solve the problems language gives rise to. However, the later-Wittgenstein rejects essentialism. Properly undertood language does not denote the essence of linguistic expressions to reveal their meaning: instead, meaning is use. Accordingly, how language functions and the uses to which linguistic expressions can be put is an ongoing affair. The clarity that now obtains is the simple recognition that meaning is use and because there are a multiplicity of uses for linguistic expressions contingent upon historical context and the changing needs of the users of a language, a philosophical method sufficient to reduce the problems of philosophy to one philosophical problem to be effectively eliminated is to be forsaken. Wittgenstein writes, '[i]nstead we now demonstrate a method, by examples; and the series of examples can be broken off. – Problems are solved (difficulties eliminated), not a *single* problem. There is not *a* philosophical method, though there are indeed methods, like different therapies.'[7]

However, if Wittgenstein's later thought is to avoid the charge of glaring inconsistency, then, like the account of language in the *Tractatus*, it must specify a perspective to be transcended and an approach to philosophical problem solving that is to be cast aside. For to assert the ongoing relevance of the later-Wittgenstein's account of language would secure the efficacy of his own type of therapy; however, to do so is to be committed to essentialism, to one method in philosophy. By his own standards, his cannot be the final word on philosophy; nevertheless, his approach silences philosophy from speaking of anything beyond itself, and thereby leaves philosophy ensnared in banal self-absorption. 'One might give the name "philosophy" to what is possible *before* all new discoveries and inventions.'[8] In the end, Wittgenstein replaces the problem-plagued and inefficient tradition of trying to justify the belief that terms have a universal and necessary meaning or essence, with a view of language that justifies the elimination of philosophical problems but leaves philosophy with nothing to talk about other than itself. Therefore, for all the talk of the revolutionary character of Wittgenstein's philosophy, he is committed to the very customary problem of justification and he satisfies it in a most commonplace manner. For like the person who feels justified in her convictions because she never ventures beyond the domestic circle, his description of philosophy comes at the cost of philosophers ceasing to dialogue with and share in the problems of the wider intellectual community.

Unfortunately, Wittgenstein's philosophy has exercised a tremendous influence over various Anglo-American philosophical movements in the contemporary era. For example, Kuhn and the sociology of knowledge[9] and Rorty's neo-pragmatism[10] have each succumbed to the conclusion that philosophy can never rise above self-description. Having rejected as characteristic of philosophy with a capital *P* both truth as correspondence and the possibility of objective representations of reality, Rorty states:

In a post-Philosophical culture, some other hope would drive us to read through the libraries, and to add new volumes to the ones we found. Presumably, it would be the hope of offering our descendants a way of describing the ways of describing we had come across – a description of the descriptions which the race has come up with so far.[11]

However, any attempt to establish the merits of one description over another ultimately terminates in a description of how a person uses the language she uses. If a person accepts a particular description in working out a private life plan then that is that. Rorty seems to think that familiarity with a diverse number of descriptions will facilitate the universal ironism that he advances as essential to a liberal utopia, but he can neither raise nor answer the question of how liberal ironism is better than any other description.[12] The end result is that philosophy fails to inform the humanities and sciences in any significant or enriching way; accordingly, philosophy is not desuetude, but a dismal consequence of lowered expectations and moribund ideals.

Continental philosophy likewise struggles with the issue of philosophical questioning. Deconstruction, for example, asks its readers to consider that traditional modes of Western philosophical questioning have favored particular dualities and emphasized a hierarchical structure that reflects fundamental prejudices, for example permanence over change, necessity over contingency, and most importantly speech over writing. The latter dualism is of great importance to deconstructionists, because on their reckoning Western philosophers have concluded that mental thought understood as interior speech has an immediacy that privileges it over writing. The intimacy a speaker has with her own thoughts has been taken by traditional Western philosophers to evidence a level of awareness indicative of certain and necessary knowledge, that is, an awareness that does not give rise to doubt. Moreover, for some propositions that make up thought-speech the immediacy of the relationship between subject and predicate, between sign and that which is signified, serve to identify the two and such identities are viewed by many Western philosophers to denote essences, self-certifying statements, or transcendental truths. Writing, viewed as an attempt to communicate this immediacy, forever falls short of its aim because subsequent discourse is distanced from its original subject as it strains to appropriate synonyms and metaphors to explicate it. The presumed identity and immediacy characteristic of thought-speech is unwittingly taken to be a standard and a foundation for justifying the elements of a particular philosophical system, while according to deconstructionists, adequate recognition is not given to the alterity/otherness that identity presupposes in its self-conception. Thus writing, viewed as the poor relative of thought-speech, is all too easily dismissed, and it is the aim of deconstruction to force philosophers to consider the primacy of writing – that thoughts flow out of it and are not merely put into it. Therefore, on the reckoning of deconstructionists, to understand the philosophy of a given thinker we have to study how her text is put together, including how the things that are not written down inform what is written. In the end, the duality of speech and writing gives rise to the contrast between necessity and contingency and truth and fiction among other central dualisms. The result is that according to deconstructionists, traditional Western philosophers have stressed a philosophy of presence, wherein notions such as truth, substance, and permanence are emphasized because they mirror the immediacy and certainty characteristic of thought-speech. Having concluded that the projects of both modernism and its intellectual progenitors cannot justify their commitment to a philosophy based upon the primacy of thought-speech, deconstructionists assert philosophy can at best uncover how the arbitrary imposition of a thought-speech-based worldview excludes and exploits those who do not share its favored dualisms and hierarchies. However, although deconstruction prides itself on questioning received philosophical verities; nevertheless, it is constrained from performing genuine philosophical work by the structure of its own analysis. Having rejected the approach to philosophy in terms of dualisms, deconstruction falls back on the core dualism of question and answer that it can neither deny nor situate itself in.[13] Michel Meyer writes, 'I am sure that Derrida would question such a view of thought, but thereby he would be questioning, and if he is questioning, his questioning would be a response, which would thereby reveal the primacy of the question–answer duality on which philosophy, any philosophy, also his own, rests.'[14] Moreover, nothing seems to invoke and to reify the philosophy of presence and its commitment to the primacy of identity more than issues of self-identity occasioned by philosophy reflecting on itself. Not unexpectedly, deconstructionists feel the need to apologize for the practice of philosophy or, worse yet, to make fun of it. Thus Pascal could be quoted approvingly, 'To make fun of philosophy, is to be a philosopher.'[15]

Chastened of all aspirations, the above contemporary philosophies are devoid of the novelty and encounter with the unknown characteristic of true romance. Domestic ennui reigns in a domicile of decidedly well-worn prospects. For such thinkers, nothing is doing in philosophy because there is little or nothing for philosophy to do other than to describe its own limitations. However, undue humility is due a comeuppance: enter Karl Popper. For Popper, philosophy is part of the ongoing rational discussion of humankind.[16] Thus what we call philosophy changes over time and thereby is a matter of convention; however, it doesn't follow that rationality is therefore an arbitrary affair. Popper asserts philosophy merits no special status as foundational and it has no distinctive subject matter or methodology, but these facts don't entail that its ability to contribute to the quest for truth is lamentable.[17] Philosophers have embellished their discipline with claims for a distinctive methodology and subject matter because foundationism begs the issue of justification, and the quest for justification has led philosophers to structure their questions accordingly and so seek a determinate methodology and content that can satisfy their justificationist demands. This commitment to justification informs what justificationist philosophers take to be legitimate questions, how they carry out their inquiries, and what responses satisfy them. Wittgenstein, Derrida, and their followers rightly take to task the traditional attempts by philosophers to justify their proposals; however, they wrongly take seriously justification as an ideal and respectively conclude that because it cannot be satisfied, philosophy can never elicit questions beyond those of self-description (Wittgenstein) or transcend a deferral of meaning that ultimately undermines the possibility of genuine inquiry (Derrida). In contrast, Popper rejects justificationism in all its forms; logic is the organon of criticism[18] and philosophers have and will continue to develop ways of being critical that arise in response to the work of both past and present philosophers, scientists, technicians, and artists. Thus for Popper, the philosopher qua philosopher can contribute to the rational conversation of humankind, where rational denotes being open to criticism, the activity of criticizing, and taking seriously the results of critical exchanges. Now this emphasis on reason as the organon of criticism and not justification is at the heart of Popper's philosophy in general and his account of objectivity in particular.

In this work, I advance the claim that Popper's philosophy provides us with a framework to answer all philosophical questions concerning objectivity: epistemological, metaphysical, ethical, political, and linguistic. Such a project risks answering questions that are alive to some philosophers but dead to others; however, to the extent that such concerns are advanced under the banner of justification they are not Popper's concerns. Indeed, it is the fact that philosophers have truck with justification at all that the objectivity of the various areas of philosophy is at issue.[19] And although the following topics are discussed in greater detail in the next chapter, it is worthwhile to acknowledge at the outset that Popper's rejection of justificationism has its source in Hume's analysis of inductive inference and his own critique of essentialism. Popper concurs with Hume that experience cannot underwrite universal generalizations of any kind. This is the case both if we accept the experiential starting point as certain and the domain of inquiry is finite. For as Hume has shown, even if we accept the data of our senses as certain, because of its restricted scope such a starting point is insufficient to establish any empirical claim that can take us beyond the testimony of memory or immediate sense experience.[20] Concerning experience in a finite universe, the claim to have inspected all of a finite universe is itself a universal statement and so question-begging.[21] Moreover, even singular observation statements such as 'This is a glass of water' are not empirically verifiable since the terms water and glass implicitly reference law-like qualities or dispositions that transcend singular observations.[22] Additionally, Popper's rejection of essentialism is to disavow the prospects for foundational claims that could conclusively establish or justify pretensions to philosophical system building of an irreproachable kind. Accordingly, Popper's rejection of essentialism is not far removed from Quine's dismissal of analytic a priori judgments. However, as we shall see, Popper's approach proceeds along methodological lines and not from an analysis of meaning.

Popper's rejection of justificationist philosophies invites consideration of his own methodological proposals, especially as they are relevant to his treatment of objectivity. For Popper:[23] (1) The aim of inquiry is the classification of statements as true or false; no statement can be established or justified as either true or false. (2) The objects of inquiry are guesses; classification proceeds by controlled guesswork in conformity with deductive logic, i.e., guesses are controlled by other guesses which are not justified but corroborated. (3) Statements are said to be corroborated that have survived the best attempts at refutation; here corroboration denotes a report on the current status of the critical discussion. (4) No justification is requisite to admit a conjecture for critical discussion, proposals do not have to prove their pedigree; however, inquirers interested in distinguishing true statements from false statements proceed by advancing conjectures that are in principle criticizable, and such conjectures can and will take into account background knowledge including prior critical exchanges. (5) In light of criticism, inquiry proceeds in reference to yet unfalsified theories; however, the classification of conjectures as true or false is understood to be fallible and thus tentative. Strict proof or disproof of a theory is not possible. This fact does not affect the quest for truth: truth and justification are distinct. (6) Understanding the nature, scope, and limits of criticism is key both to distinguishing critical thinking from dogmatic thinking and to identifying an appropriate role for tenacity in theoretical and practical endeavors.

On Popper's reckoning, his anti-justificationist deductivist philosophy advances an approach to problem solving that begins by recognizing the value of commonsense realism. The growth in knowledge characteristic of the humanities and sciences is a consequence of critically engaging what commonsense tells us to be true. Science is *'common-sense knowledge writ large'*,[24] and the growth of knowledge beyond the commonsense level is best studied by advancing bold conjectures about the world structured in a manner that exposes them to criticism. Although an inquiry into the growth of knowledge sounds like the most staid endeavor characteristic of traditional epistemology, Popper's treatment revolutionizes the problem of knowledge because it leads us to acknowledge that science, the greatest source of our knowledge about the world, is more often than not everything epistemologists say it should not be: 'unjustified, untrue, unbelief'.[25]

However, the claim that Popper's philosophy contains the framework for a comprehensive account of objectivity able to satisfy the various areas of philosophy seems to fly in the face of his own philosophical prescriptions. Popper recognizes that ubiquity and triviality are not unrelated. For example, he argues Freudians could not only explain all of human behavior, but could also explain away any criticisms of psychoanalysis by appealing to facts within psychoanalytic theory. For Freudians, no aspect of human affairs is immune to Freudian analysis; consequently, Popper argues Freudian analysis is itself immune to criticism, hence dogmatic and thereby trivial.[26] Philosophers want the truth in every domain, but theories structured in a doctrinaire manner to secure the truth everywhere betray the openness requisite for truth to be a consequence of rational inquiry. Even a true theory if it is structured and held in a manner that makes it immune to criticism undermines rational inquiry; indeed, for Popper, a smothering ubiquity is indicative of authoritarian aims and entrenched power seeking to retain its prerogatives.[27] In contrast, Popper's approach to problem solving involves a distinction between a true theory and a theory known to be true. How a theory is structured to be an object of inquiry concerns the conditions by which its truth is ascertained. The great mistake characteristic of Western philosophy is the tendency to conflate truth with justification, since the received view is that a theory can only be cognized as true when it has *shown* itself to be true, that is, it is justified by having satisfied a criterion of truth that is identified in light of how we acquire evidence for the theory at issue. Thus empiricists seek an empirical criterion, and rationalists demand an a priori criterion for truth, etc. This practice of correlating the manner truth is discovered with a criterion of truth is motivated by a fear of error that has the consequence of thinkers seeking to justify their theories by means of some unwavering foundation.[28] Today, nearly every philosopher is a fallibilist to some extent, having realized the less than alluring prospects for justification of a conclusive kind.[29] Nevertheless, contemporary philosophers somehow feel entitled to draw conclusions concerning the demise of a standard that never should have taken the field in the first place. This is akin to an instructor advancing levitation as a requisite for passing her course and drawing conclusions about her student's motivation, intelligence, and industry when her illegitimate requirement is not met. Because justificationist philosophers have taken truth and justification to be coextensive, truth is the first casualty of the attacks leveled at justification: witness the denial of objective truth by Wittgensteinians, deconstructionists, and post-structuralists *et al.*

Popper is the first to see things rightly. The goal is not to justify our theories as conclusively or even probably true; rather, the aim is to make 'fallible knowledge claims objective'[30] as part of the quest for truth. Accordingly, philosophers are permitted to return to the traditional topics of philosophy with an eye to classifying our theories (i.e., system of statements) as true or false rather than having an eye to establishing them as such. The aim and process of classifying theories as true or false demands that we first acknowledge them to be guesses; by doing so we avoid the temptation to treat them as indicative statements and so pursue other indicative claims as an intellectual abutment. Thus not only are inductivist and foundationist tendencies precluded, but so too is the system building that leads to absolutism. Throughout his career, Popper excoriates each of these philosophical transgressions through a variety of critical approaches as part of working out his own non-foundationist, anti-authoritarian, indeterminist, realist, and objectivist worldview. Popper works out this worldview first and foremost by exploring the merits of bold conjectures and intersubjective criticism in the natural sciences and politics, and in doing so he thereby reveals that fallibilism and objectivity can be reconciled in a way that promotes a framework for the resolution of actual political and scientific problems. Judgments of success concerning proposed scientific and political solutions are tentative, since today's victory can contain the seeds of tomorrow's defeat, but this fact shouldn't be exploited in the service of a sociology of knowledge that would deny objective truth by making truth relative to historical and social contexts. Instead, the temporally conditioned nature of our judgments of success suffices to indicate the fallible nature of our scientific and political practices and hence the need for non-foundationist critical methods that can, when necessary, themselves be taken to task.

The fecundity of a philosophy is evidenced as much by the questions it raises as by the questions it answers. Popper and his followers ask not only that humankind take appropriate pride in their hard-won intellectual victories – we should desist from becoming a generation 'too mentally modest to believe in the multiplication tables'[31] – but that we look hard at the material and formal elements of criticism in attaining those victories. To this end, Popper's student Bartley in his groundbreaking work *The*

Retreat to Commitment draws an important distinction between justificational and non-justificational philosophies of criticism. The former is based upon a 'fusion' of justification and criticism and comprises a strong and a weak type.[32] The strong version advances the notion that 'the rational way to criticize an idea is to see whether or not it can be rationally justified'.[33] Bartley clarifies this point using as an example Hume's empirical criterion of meaning. Hume holds that an idea is meaningful and avoids 'sophistry and illusion' only if it can be grounded in its corresponding sense impression(s). Thus, Hume concludes that the procedure of reducing a simple or complex idea to its sense impression(s) comprises *the* standard of rationality and the legitimacy of an idea is to be evaluated in reference to how it meets this standard. The weak type of justificational criticism doesn't require a theoretical or practical claim to be derived from the standard but only that it does not conflict with it. For example, the United States Constitution is the standard for justified legal behavior in the United States, and any action is permitted so long as it does not conflict with what the law proscribes.[34] The multifarious and complex nature of human behavior precludes the Constitution from prescribing all possible permissible actions for United States citizens, and so the Constitution's negative constraints function as a weak form of justificational criticism concerning the actions citizens may undertake. In the end, Bartley argues each type of justificational criticism is unable to justify its own basis in a non-question-begging way. Hume, for example, can't justify his empirical criterion of meaning without inviting an infinite regress; consequently, he asserts sense data as the irreproachable basis for his empiricist philosophy. Thus, Bartley asserts justificationist philosophies invite a *tu quoque* of the most embarrassing kind, and thereby engender a crisis in rationalist integrity. His point is that the justificationist, believing herself to be a proponent of the use of reason to adjudicate intellectual disputes, challenges spurious intellectual claims (e.g., phrenology, creation science, astrology), by demanding that the advocates of such ideas justify their theories. However, the justificationist cannot satisfy her own requirement without inviting an infinite regress or advancing a foundation that need not be justified, thus failing to satisfy her own justificationist demands. Accordingly, Bartley concludes that all justificationist philosophies resolve themselves to authoritarian dictates. Indeed, Bartley argues that an authoritarian basis is found in Popper's early formulation of critical rationalism. Popper's critical rationalism asserts that all theories are to be advanced in a manner that exposes them to criticism. However, Popper argues the use of reason in its critical mode is based upon 'an irrational *faith in reason*'[35] and thus he posits a limited fideism as a type of irrational basis for the use of reason. Bartley amends Popper's critical rationalism in a manner that is consistent with Popper's own anti-justificationist and fallibilist tendencies producing comprehensively critical rationalism (hereafter CCR), which holds that 'a position may be held rationally without needing justification at all – *provided that it can be held open to criticism and survives severe examination*'.[36] Bartley's CCR is understood by many to have been adopted by Popper as part of his philosophy,[37] and like all good philosophy it has engendered a host of questions and quandaries that have kept philosophers busily employed since it was first proffered. Central among these issues is whether CCR is itself criticizable and thus satisfies its own demands. This debate, which will not be detailed here, rightly highlights fallibilism as a seminal component of Popper's critical philosophy. In fact, the relationship between criticizable theories and fallibilism has been misunderstood to the detriment of a proper appreciation of the truly revolutionary character of Popper's thought.

The need for critical methods to control the error-laden and hence fallible nature of our theories is recognized to a greater or lesser degree by various intellectual traditions, and the emphasis on the criticizability of our conjectures is an obvious consequence of fallibilism; yet what is not duly appreciated is that the criticizable nature of a theory does *not* entail that it is possibly false.[38] Indeed, the assertion that criticizability implies possible falsity has resulted in the tendency to hold uncritically its contrapositive: if a theory is not possibly false (i.e., it's true), then it is uncriticizable. Wanting to endow their theoretical and practical claims with as much truth as possible, Western philosophers have been motivated to seek methods of securing the truth that make their theories free from falsity and thus on their reckoning their ideas are established as true if no criticism can be lodged against them. What is at issue here is not the well-known fact that people can and do adopt system-saving stratagies so to immunize their theories from criticism; rather, the issue is that system-saving stratagem are a consequence of having erroneously identified the criticizability of a theory with its possible falsity. Once this identification is made, the natural and obvious temptation is to seek self-certifying statements, transcendental truths, and the like, as foundational claims that are viewed as obviously irreproachable or to advance a standard of justificational criticism that takes as its basis something that is held aloof from criticism because it is understood to be apiece with reason. What Popperians appreciate to a degree that their philosophical counterparts do not is that a growth in our awareness of the truth[39] can occur not only by deducing true consequences from a true theory, and sometimes by false premises entailing a true conclusion, but even by criticizing true theories while simultaneously acknowledging their truth. The fact that the immediate response of most thinkers is to ask what is to be gained by criticizing true statements is evidence that they do not appreciate the merits of criticism dissociated from a justificationist and authoritarian agenda. The idea has been that the truth of a claim is its own justification and the basis for a theory's authority; accordingly, criticism, whose aim it is to expose error or falsity, has no work to perform when truth is present. However, consider the difference between the following two approaches to the existential judgment 'I exist'. The statement 'I exist' when Descartes conceives it is understood by him to be so clear and distinct that its truth is its own justification. Moreover, the truth of this statement becomes the self-evident and authoritative basis for an entire philosophy. However, from the perspective of Neo-Darwinian evolutionary theory the judgment 'I exist' is not justified by any state of affairs; rather it is a function of a person's relationship to the environment and her genetic makeup. The environment is a critical check on how an organism/person relates to it; that a person exists is not a justification or an assurance of her current survival, and current survival is what 'I exist' means in a biological context. To understand existence from a biological perspective, the best a person can do is to consider critically her survival-based behaviors and biological structures to understand better the truth of her existence; however, her behaviors and structures are not a *guarantee* of existence in the same way that Descartes takes his act of conceiving his existence to establish the indubitability of the judgment 'I exist'. Starting from the truth of her personal existence, and without ever trying to justify what she takes to be true, an individual can enormously enrich her self-understanding by critically engaging her presence in the world by means of evolutionary theory and its attendant disciplines of ethnography, cell biology, ecology, etc.

Genius-inspired insights and their accompanying revolutionary changes in a discipline often involve getting its practitioners to think critically about what is taken to be true. Darwin did this concerning the existence of human beings, and Cantor's development of set theory did the same concerning our intuitions about number. Admittedly, in the Western intellectual tradition criticism has predominantly taken the form of justificational criticism (this is the case in Darwin's writings as well), but this fact doesn't affect the logical situation that such criticism exposes itself to a *tu quoque* and leads to the retreat to ultimate commitment, and thus the crisis of rationalist integrity, initially recognized by Popper and responded to in a more comprehensive way by Bartley and Miller.[40] Popper's non-justificationist philosophy of criticism lays bare the full scope of criticism in the service of reason; it avoids the quagmire of justificationism that, especially since the Renaissance, has led philosophers both to adopt skepticism as an epistemic state and to develop philosophical positions as a foil to skepticism as an epistemological option. The fallibility of our cognitive apparatuses, of our theories, and of our technological enterprises led philosophers to advance skepticism as an epistemological position primarily because, taking beliefs as the object of inquiry, they failed to find a basis for unremitting consent. However, Popper's philosophy welcomes skepticism as a *methodological* tool essential to rooting out error.[41] The distinction between skepticism as an epistemic state and as a methodological tool is of fundamental importance. When adopted as an epistemic state, skepticism takes seriously the standard analysis of knowledge and the quest for knowledge in terms of justificationist priorities. On methodological grounds skepticism acknowledges our fallibility, but in its methodological role it doesn't take truth to be something to be appended to beliefs and thereby transforms them into certain knowledge. If taken to denote an epistemic state, it is not our inability to attain certainty that is worrisome to Popperians but its trivial omnipresence. Certainty as an epistemic state can be established by any means when it is taken as the aim of inquiry intended to replace doubt, including the radical doubt of the skeptic, understood as another epistemic state. If certainty is the aim, and if flipping a coin induces certainty as an epistemic state, then there is nothing more to be said about the matter, since there will be no need to consider additional evidence whether confirming or disconfirming.[42] Accordingly, Popper approves of the linguistic turn in philosophy insofar as it dissociates questions of epistemology from the pursuit of mere epistemic states such as beliefs. Moreover, on his reckoning, understanding our theories as systems of statements is also enhanced by Tarski's theory of objective truth that defines truth in reference to material adequacy and formal correctness understood as objective properties of any language in which a definition of truth can be formulated.[43]

The tendency to equate criticizability with the possible falsity of theories and so seek to construct or identify theoretical and practical claims that are beyond criticism has its origin in an unhealthy approach to error. Miller, the principal expositor of Popper's critical rationalism and a formidable philosopher in his own right, reminds us that it is not the perpetration of error but its perpetuation that we should seek to avoid.[44] If inquirers fear error to the point of not wanting to advance any theoretical or practical claims, that is, they adopt a better safe than sorry policy, then the growth of knowledge is forestalled because nothing is ventured and so nothing gained. 'Our enchantment with truth can only be unsettled by the specter of error if error is beyond any effective control and so forever haunts our theories.'[45] However, on formal grounds alone it seems the perpetuation of error is indeed irrevocable and cannot be exorcised. The ability of Popper's philosophy to address this issue is of such importance that consideration of an extreme case of how error perpetuation *seems* irrevocable must be considered as a backdrop for explaining the general features of his defense of objectivity. Much of what follows advances the same conclusion as Miller and Tichy's independent arguments against approximation to the truth; but rather than emphasize the problem of a proper metric as they do I treat the problem of an unchecked growth in error.[46]

To understand how the perpetuation of error is logically possible we need only consider a variant of the 'tacking-on' thesis.[47] By the valid inference rule Addition: $p/\therefore p \vee q$, any statement may have added to it any content whatsoever. For example, by Addition the following disjunctive conclusion is a valid inference from the given premise: My desk chair weighs 50lbs. Therefore, My desk chair weighs 50lbs or God is in heaven. If we accept the premise as true then the resulting disjunction is true and this can be determined by a simple truth table.

Now Popper's methodology of conjectures and refutations proposes that we get to the truth in three ways: (1) by conjecturing true statements that can be classified as true, (2) by eliminating from a set of competing theories comprising both true and false elements those theories that are false, and (3) by eliminating the false consequences of a theory in the hope that what remains is the truth requisite to address the problem situation that first initiated our conjectures. Lacking a systematic way to generate true conjectures, Popper argues the rational option for inquirers into the truth is to focus on error elimination in the sense of (2) and (3) above. However the fallibility of our theories is not insignificant. Indeed, our background knowledge concerning the history of thought reveals that the vast majority of our theoretical and practical claims have shown themselves to be false. If such error is able to perpetuate itself unchecked, then the hope of arriving at the truth by critically engaging our theories in the name of error elimination is fundamentally misplaced. Again, consider the tacking-on thesis this time in conjunction with the independently obtained result of Keuth and Vetter that every false theory has the same number of true consequences as false consequences whether the set of consequences is finite or infinite.[48] Now the Addition rule designates that false content can be added to the false consequences of a false theory: Let L designate Lamarck's theory of evolution, and let W be the erroneous consequence that water birds developed the adaptation of webbed feet by continuous stretching of the skin on each foot during swimming. Given the Addition rule, it is logically possible to increase the content of this false consequence by adding to it a false claim X from biology or any other discipline: $W/\therefore W \vee X$. On logical grounds error grows in perpetuity.

If correct, the above argument has the following problematic consequences: (1) Every false theory understood to have a finite number of false consequences logically allows for an infinite number of false consequences; accordingly, error perpetuation is irrevocable and grows in a deductive manner. (2) Given the result of Keuth/Vetter and keeping in mind that the above example deals with a false theory, the number of true consequences grow in a one-to-one correspondence with the number of false consequences; consequently, a growth in error results in a growth in truth content. (3) There is no purely logical reason to prohibit the growth of false content since it is a well-known fact that false statements can validly imply true conclusions. At times we are happy that errors arise (sometimes in the form of accidents) since they can lead to the discovery of new truths.

2. Overview of Popper's Solution to the Problem of Objectivity

Popper's philosophy is extremely attentive to the formal conditions requisite for criticism to be possible. Central among these is the demand that theories not be self-contradictory. In the case of empirical science Popper argues

> the importance of the requirement of consistency will be appreciated if one realizes that a self-contradictory system is uninformative. It is so because any conclusion we please can be derived from it. Thus no statement is singled out, either as incompatible or as derivable, since all are derivable....This is why consistency is the most general requirement for a system, whether empirical or non-empirical, if it is to be of any use at all.[49]

Moreover, in the same passage Popper points out that it is not the falsity of a self-contradictory theory that is problematic, since we often work with false theories while simultaneously trying to improve them;[50] instead, it is because contradictions fail to preclude any state of affairs that they aren't testable and thereby are opposed to rational inquiry. In the same way, the tacking-on thesis as a purely formal consideration fails to specify a determinate problem situation; thus, it proposes to compound error in a way that isn't amenable to error elimination. For Popper a genuine conjecture has to specify prospects for the possible elimination of itself as well as its consequences. To the extent the tacking-on thesis fails to specify a determinate problem situation it provides nothing for criticism to respond to.[51] Kant's famous dictum can be profitably modified: Criticism without a genuine problem situation is empty; a problem situation that can't specify conditions for criticism is blind – to the prospects of rational inquiry.

With the identification of a problem comes the identification of modes of attacking it. Thus, concerning natural science, Popper asserts experience is the method by which we can increase our understanding of the one world we call our own.[52] The coupling of experience as a critical methodology with fallibilism provides the young Popper with the basic elements of his theory of objectivity. To acknowledge error is immediately to undermine rudimentary forms of idealism and solipsism in the sense of reducing reality to some aspect of the knowing subject. Simply put, if reality is our construction, then why would we so often be wrong concerning our speculative and practical claims? Also, the fact that we have to guess concerning how things are in the world, and the world sometimes shows our guesses to be in error, indicates that reality is recalcitrant in a manner that is not of our own making. This is not to say that there is no role for the knower and no a priori element in theory construction. Like Kant, Popper holds that experience only responds to questions of our own asking, thus our theories are not built up from observations in an inductive fashion by direct reference to experience, but imposed a priori upon a problem situation suggested by experience. For Popper it is modern science that best explains how experience suggests a problem situation to us. Popper argues the recognition of a problem situation is first and foremost the result of genetically based expectations about the environment that are the consequence of evolutionary adaptations. For example, that humans perceive light waves at the frequency of 10,000mhz fixes the range of expectations concerning visual reality and thus determines what we call visual experience. With a little charity we can understand these expectations to function like theories about the environment. However, Popper pushes the evolutionary construct one step further by asserting the expectation for regularity and hence lawlikeness is the key adaptation to an increase in knowledge about the world. Thus, like Kant, Popper holds that the knowing agent plays the fundamental role in determining reality since the expectation for lawlikeness is genetically a priori; however, unlike Kant he denies that any particular law or theory or expectation is a priori valid. For example, the notions of space and causality that underwrite the Newtonian worldview that Kant accepts have been undermined by developments in contemporary physics. Consequently, for Popper, our a priori expectations are no more immune to criticism than the unexamined prejudices and beliefs that have been the target of philosophical inquiry since the time of Socrates. Agassi points out that Popper's philosophy nicely navigates between the Scylla of nature and the Charybdis of convention that plagued antiquity. Popper rejects this dichotomy by making a place for a conventional element within scientific theorizing and scientific method while acknowledging that scientific truths 'are not truths by nature, yet they are not arbitrary either'.[53]

Sensitivity to our own shortcomings, coupled with the aim of classifying as true those theories that to date have not shown themselves to be false, requires that we make our theories public and thus expose them to intersubjective criticism. Moreover, our insensitivity to our own shortcomings makes the demand for publicity a good policy to adopt, because we can expect others to be critical of our claims if we fail to do so. Thus publicity, and in the case of empirical science repeatability, which goes hand in hand with the public character of science, are essential to Popper's deductive theory of testing in the service of truth. For Popper, the requirement publicly to test our conjectures is best supplemented by a move away from belief epistemologies as discussed above and thus he posits his much maligned World Three, which is a partly autonomous and publicly accessible realm of objective statements that is the basis for self-transcendence.[54]

The public nature of criticism demands a public context where criticism can flourish, and therefore a political arena that eschews authoritarianism, dogmatism, and provides the political means, even if the political will is lacking, to eliminate policies that prohibit free critical exchange. For Popper, such a public space is an 'open society' and is characterized by piecemeal social engineering, that is, the procedure of tentative trial solutions and error elimination on a localized level, in opposition to utopian social engineering that mandates holistic social policies that would hand humankind happiness in exchange for freedom and critical rationality.[55] This political arena must be coupled with standards of personal ethics: humility, openmindedness, honesty, and courage, which are not a pre-critical basis for criticism as Popper seems to imply, but instead are the logical consequence of identifying rationality with criticism. This claim, although not found in Popper's writings, is the key to solving the problem of the relationship between reason and morality often phrased 'Why be moral?' and is addressed in the final chapter of this monograph.

Thus for Popper, morality, politics, and of course science are all elements in a rational theory of tradition that does not capitulate to authoritarianism or relativism, but places objectivity at its core. As components of a tradition each of these three are equally relevant, although Popper's writings reveal that he is most concerned with issues of politics and science. Concerning the latter, Popper will argue for both indeterminism in physics and the selective nature of the environment in evolutionary biology as features of the real world that compliment his account of scientific method and critical rationality. Both indeterminism in physics and natural selection in biology entail a real world that is recalcitrant, and thus displays a novelty that in the case of humans promotes both humility and courage concerning the unknown and the unexpected. In the case of biology, the feedback that organisms get from the environment may lead them to modify their behaviors as a response to novel situations. Thus Popper argues Darwinian evolution by natural selection simulates Lamarckian evolution by instruction.[56] Accordingly, Popper's philosophy holds open the prospect of improving the human condition through criticism, and this prospect begins on the biological level. Thus Popper as the originator of evolutionary epistemology advances one of the most important contemporary insights towards understanding the growth of knowledge. On the purely biological level non-human organisms avoid criticism or selection from the environment at all cost. In contrast, humans have shown themselves to be different because, in ideal circumstances anyway, they consciously seek to engage their ideas critically and even go so far as to set up traditions and institutions that promote critical inquiry. The fact that humans promote critical inquiry is the principal basis for Popper's insightful contrast between Einstein and the amoeba. 'Einstein consciously seeks for error elimination. He tries to kill his theories: he is *consciously critical* of his theories which, for this reason, he tries to *formulate* sharply rather than vaguely. But the amoeba cannot be critical *vis-à-vis* its expectations or hypotheses; it cannot be critical because it cannot *face* its hypotheses: they are part of it.'[57] In emphasizing the biological basis to criticism Popper is not merely advocating a type of reductionism in the same fashion that advocates of the sociology of knowledge assert that all knowledge claims are socially conditioned. The pursuit of reductions, like skepticism, has methodological merit. Reductions can be informative, and when the elements that exposed in a reduction are clearly identified it can be a springboard to critical inquiry as we explore how those elements produce more complex phenomena. However, failure to acknowledge the methodological aspect of reductionism results in a kind of essentialism; since the ideal reduction is a definition, the identity of definiendum and definiens. In contrast to the identity between definiens and definiendum, one important merit of an evolutionary epistemology is that it teaches us that the fit or identity between an organism's expectations/theories and the environment is never secure. The cockroach, although having survived as a species for 140 million years, is never entitled to be cocksure.

The fallibility that belongs to humans as organisms mandates a humility that isn't always captured in Popper's account of his methodology as conjectures and refutations or in biological analogues invoking the fitness of a theory with the world. A more benign rendering is 'I may be wrong and you may be right, and by an effort, we may get nearer to the truth.'[58]

Intersubjective criticism, the recalcitrance of the world as present in physics (indeterminism) and biology (natural selection), methodological skepticism and reductionism, the conjectural nature of knowledge, the move beyond belief epistemologies to a treatment of theories as systems of statements regulated by the correspondence theory of truth and as expressing objective propensities are the fundamental features of Popper's treatment of objectivity. To appreciate how these elements come together in a comprehensive way I next consider the problems they are meant to address.

3. Comments on the Problem Situation

In the above, I make much of Popper's insight concerning the importance of having a criticizable problem situation as an aid to rational inquiry. In what follows I want to present a brief overview of what the problem of objectivity is in some of the traditional divisions within philosophy. Disclaimers concerning the possibility of a complete account for each respective philosophical discipline are in full force here. A full description of any one of these problems merits a book of its own. What I hope to do is capture something of their traditional flavor and then indicate how they must change in reference to Popper's philosophy. Consequently, Popper's philosophy demands that other philosophers seriously consider that they have been asking the wrong questions all along.

At this stage of the inquiry it should come as no surprise, although it is nevertheless a surprising innovation, that Popper's philosophy is going to force philosophers to ask, 'Within the traditional divisions of philosophy how must the discipline be structured if criticism is to be possible?' That the respective answers invoke aspects of the account of objectivity identified above should be expected. Moreover, the parallels between this question and Kant's general approach to philosophy should not be ignored, but neither should they be overemphasized. Popper asks us to consider that his philosophy 'puts the finishing touch to Kant's own critical philosophy'.[59] In contrast to Kant, who requests the conditions for the possibility of knowledge as a science and thus for a transcendental foundation that is a priori and hence universal and necessary in scope, Popper recognizes that each philosophical discipline and each problem of life must be able to specify how it can expose itself to criticism. Problems in philosophy and in life in general don't fall into pre-made rubrics or categories, but are homespun; accordingly, the procedures by which we critically engage them must be equally homey.

3.1. The Epistemological Problem

The nature of knowledge, like so many other philosophical topics, is disadvantaged by the metaphors employed to explicate it. Most often physical metaphors are used that reference grasping or seeing as aids to help us cognize what we take to be most characteristic of knowledge. For example, 'I see what you're saying' or 'I grasp your point'. Implicit in such metaphors are the ideas that knowledge is a property or possession of a knower, and a union between knower and thing known. Because of their subjectivist presuppositions, Popper rejects both the idea of knowledge as possession of a knower and the idea of knowledge as a unique type of union. Central to both notions is the priority they give to subjective states. Traditionally, epistemologists have asserted belief to be the epistemic state possessed by a knowing agent that is transformed into knowledge by the addition of truth and justification. However, as already noted, if we take certain belief as the aim of inquiry, there is no need to consider additional evidence whether confirming or disconfirming. Moreover, when a belief attains the status of conviction the intimacy a person has with her own belief states lends them an authority that engenders dogmatism, prohibits critical thinking, and can even lead a person to kill for the sake of what she believes.

Given the physical metaphors that describe our commonsense-based approach to knowledge, in the same way that we can put to our senses the question 'What is it?' concerning the objects of sensation, by introspection philosophers have treated knowledge as an object of inquiry and have raised the same essentialist query. According to the traditional analysis of knowledge, the answer to the question 'What is knowledge?' is justified true belief. The fact that these elements are taken to be necessary and jointly sufficient for knowledge has led to the problems of defending the objectivity of each element of this tripartite account, or of supplementing it in some way if the tripartite account is found wanting. Philosophers such as Dewey and Wittgenstein have attempted to shift the question away from an essentialist formulation by asking that we reflect upon the uses of knowledge and how it is that the term knowledge itself is used. If we adopt the later-Wittgensteinian perspective, one use of the term 'knowledge' common to academic and non-academic discourse alike is for it to function as an indicator of success in reference to socially defined standards: 'Knowledge' is a success word. This approach fundamentally differs from the essentialist analysis of knowledge by rejecting both the essentialist's implied account of consciousness as an arena where knowledge is located as a special object of inquiry, and the treatment of knowledge as a unique type of representation. However, both the essentialist and meaning-as-use based approaches to the problem of knowledge are, on Popper's reckoning, deficient to address the problem of the growth of knowledge and its attendant problem of demarcation. It is these two interrelated issues that are the focal points of Popper's epistemology. 'The central problem of epistemology has always been and still is the problem of the growth of knowledge.'[60] And in *The Logic of Scientific Discovery*, Popper tackles as part of his focus on the growth of scientific knowledge the problem of demarcating empirical science from non-scientific undertakings, especially metaphysics. Later in his career Popper will expand his philosophical outlook and emphasize the distinction between justificationist approaches to the theory of knowledge and his own critical approach.[61] The essentialist treatment of knowledge attempts to demarcate knowledge from opinion, true belief, etc., and the meaning-as-use approach to epistemology must minimally indicate why the term knowledge is appropriately used in one context and not in another. Thus in a fundamental way both approaches must ask how it is that a knowing agent can establish her knowledge claims as either representations or as having a particular use. For those philosophers whose analysis of knowledge is a response to the essentialist question, justification has played the central role in demarcating knowledge from other epistemic states. It is clear that Popper unequivocally rejects any appeal to justification as a principle of demarcation and as contributing to an understanding of the growth of knowledge. Concerning those thinkers that advocate the meaning-as-use approach to knowledge there can be little doubt that they are interested in making distinctions concerning the use of words, but, having described the different ways that words are used, can their descriptions inform us as to why one description is better than another? Consequently, the meaning-as-use treatment of language may help a person to view the linguistic landscape, but such an understanding is at best a preliminary concerning how to cross it, since this often involves decisions as to why one path is better than another. In contrast, Popper's philosophy advances a different epistemological insight concerning the growth of knowledge in general and an increase in our understanding of the physical world in particular. Via his account of corroboration Popper contends that a preference between theories can be established; moreover, it can be established in a way that clarifies the relationship between critical and dogmatic thinking. Thus humankind can attain new truths in a non-authoritarian manner. Key to doing so is to relinquish the notions of knowledge as a possession, since pride of possession often overwhelms a person, making her hesitant to criticize something as intimate as her own convictions; moreover, the idea of a union between knower and thing known emphasizes belief states as the special vehicle for this union and invites the question of how one's beliefs are justified.

3.2. The Metaphysical Problem

As is the case with so much else in philosophy, a return to Plato's thought is most instructive. Plato offers us a sophisticated but nonetheless uncomplicated account of objectivity. The essence of his position is simple: the nature of knowing reflects the nature of the object known.[62] Throughout those writings where epistemological and metaphysical issues are prominent, Plato consistently correlates the nature of knowing with its object. The latter is immaterial, unchanging, and complete and so too is knowledge as distinct from right opinion and other less desirable epistemic states. Although original, Plato's correlation between the nature of knowing and the nature of the object known is not a singular occurrence in the history of philosophy. Medieval philosophers who make the object of knowledge the eternal and unchanging God hold the same position. Moreover, it is present in Descartes' cogito as a response to skepticism; it can be found in Husserl's eidetic reduction; and the advocates of the linguistic turn advance the same position, but philosophers of language more often than not have argued that an appropriate philosophical understanding of language undermines the possibility of objectivity. Central to the traditional metaphysical problem is the correlation between claims of objectivity and an object whose unique properties are there to be discovered, classified, put into the service of knowledge as a standard and metric, and exploited by technē. Principal among the instances of technē are: language, works of art, and the instruments that extend the operation of our senses. Obviously for Popper, to the extent the metaphysical problem has been conflated with issues of justification and foundationism it has pursed the wrong agenda. Instead, Popper's version of the metaphysical question of objectivity asks, 'What must reality be like if criticism is to be at all possible?' Because Popper's critical methodology emphasizes the elimination of falsity in the pursuit of truth, the status of false statements comes to have a prominence that must be reckoned with. Traditionally, false statements have been associated with non-being or privation of some sort, and so philosophers have struggled with how false statements inform our understanding of reality. For Popper, our mistakes and errors have a causal efficacy, that is, they affect the quest for truth and influence the path of inquiry: the truth hurts but error can be a sting that kills. Accordingly, what we require of false theories is a way to test for and to control this causal efficacy while never making our theories immune to criticism. Via his propensity theory of causality and probability, Popper's philosophy provides this in a way that links empiricism and metaphysics. Moreover, Popper argues metaphysics can function as a framework for empirical and mathematical research all the while manifesting internal conditions that show it to be criticizable. For Popper, the best way to discern the possibility, nature, and scope of metaphysical claims is to approach them in relation to the real world problems that invoke them. Concerning natural science, for example, Popper argues that it is the metaphysical problem of induction that is key to his treatment of scientific method, and it is within science that he finds a solution to it. Accordingly, metaphysical claims can't be gratuitous; rather, they must be constrained by the state of affairs that gives rise to them. No doubt, this assertion is contrary to a core element of Western metaphysics. Commonly, metaphysical claims are understood to transcend experience concerning their subject matter, scope, and justification. Minimally, the idea is something like the following: pure reason exercising the highest degree of abstraction suffices to merge both the context of discovery and the context of justification. Because metaphysical reflection is a priori cognition, its concepts are simultaneously justified as universal and necessary by being clearly conceived and the clarity of this conception is often attributed to the nature of the object cognized. For the metaphysician, the intellectual energy expended to reach the level of abstraction characteristic of metaphysical reflection is rewarded by the efficiency of uniting how one comes to know and the vindication for the knowledge attained. Moreover, in metaphysics there is a direct correlation between the identification of a priori concepts, reasoned constraints on such concepts, and the limits of conceivability. However, reflecting on the growth of human knowledge Popper argues rational inquiry as critical activity extends the scope of the conceivable. His researches focus on the role of natural science in the growth of knowledge and the manner natural science has been aided by metaphysical ideas while contributing to novel metaphysical insights. Concerning the latter, quantum physics has forced philosophers to rethink their understanding of such topics as causality, the law of non-contradiction and – an issue Popper is particularly interested in – the debate over determinism. Consequently, Popper calls himself 'a tottering old metaphysician'.[63] But Popper's metaphysics is markedly distinct from the mainstream Western tradition exemplified by Plato and understood as a footnote to it. In *The Open Society and Its Enemies*, Popper argues that Plato's metaphysics aims at arresting change so to secure a metaphysics of permanence that can underwrite authoritarian ideals. Metaphysics in the Platonic tradition thus advances an essentialism that keeps both thoughts and people in their respective places; witness the *Republic*'s hierarchical structuring of both the polis, and in the 'Divided Line' allegory, of reality and its corresponding epistemic states. In contrast, Popper argues that reflections on the Greek contribution to science, and thus on the development of a critical methodology, invites a methodological nominalism, i.e., the definiens of a given term is simply a tentative designation that serves to facilitate critical inquiry.[64] Critical inquiry requires the inquirer to interact with the world and this interaction gives rise to new metaphysical issues and insights. Therefore, for Popper important metaphysical ideas can be said to 'coat-tail' on the novel scientific developments that give rise to them.[65] Thus, there is, in contrast to the staid essentialist tendencies of Western metaphysics, something new under the sun.[66]

3.3. The Linguistic Problem

'Language not having existed, man had to invent it.'[67] This introductory sentence from a well-known text on the philosophy of language couldn't express a viewpoint that differs more radically from Popper's mature views on language. According to Popper, humans did not invent language, rather language led to the 'invention' of humans. If language minimally involves the communication of information, then language begins on the most fundamental of biological levels. For example, consider the transmission of information that takes place on the microscopic level between various parts of the cell. For Popper, to divorce an analysis of language from its biological origins is to fail to comprehend both the use of language and its significance as representation in higher organisms. In an attempt to secure the importance of Darwinism for the type of philosophical pragmatism that he espouses, Rorty writes:

> The link between Darwinism and pragmatism is clearest if one asks oneself the following question: At what point in biological evolution did organisms stop just coping with reality and start representing it? To pose the riddle is to suggest the answer: Maybe they never *did* start representing it. Maybe the whole idea of mental representation was just an uncashable and unfruitful metaphor.[68]

To advance this dichotomy between representation and use is to completely misunderstand what it is that evolutionary theory has to offer philosophers concerning the nature of language. Fundamentally, an evolutionary approach to language means that we understand human language as a homologous adaptation that parallels linguistic behavior in lower organisms. Popper, under the influence of Karl Buhler, asserts that language has a fourfold function: expressive, signaling, descriptive, and argumentative.[69]

The latter two functions are characteristic of human language and are essential to critical thinking. This hierarchical distinction concerning the function of language obviously emphasizes its use, but the distinction, for example, between different signaling techniques among organisms requires that we reflect upon how it is they represent the world, since the use of signaling as signaling is the same for each. Thus giving a representational role to language is the chance to drop the idiosyncrasies characteristic of a philosophy of know thyself as the only framework for understanding representation. By acknowledging other organisms as knowers having modes of representation different than our own, we come to see our own manner of representing as just one mode among many, and thus, the evolutionary worldview forces us to relinquish the notion of an ideal mode of representation. Moreover, to the extent our manner of representing the world is an adaptation, it has survival value and definitely a use. From a biological perspective the answer to Rorty's question is a 'both/and' response: language *is* the principal vehicle of representation, and exploration of its use must first be understood in reference to its role as an evolutionary adaptation. Popper's approach to language via evolutionary theory raises two important issues of demarcation: (1) under what conditions is it correct to assert the relevance of divergent conceptual schemes or ways of representing the world and (2) under what conditions must we acknowledge that acknowledging conceptual schemes as distinct from their referential content is at all intellectually relevant? That the previous sentence is decidedly Davidsonian is intended. Davidson, in his classic paper 'On the Very Idea of a Conceptual Scheme',[70] argues that to be a language is to be translatable. To recognize an activity as linguistic is to interpret the behavior of another in reference to one's own linguistic behavior. Thus for Davidson, insofar as behavior is designated linguistic behavior, it isn't possible to recognize it as incorporating a conceptual scheme so radically different than one's own such that it isn't amenable to ascriptions of meanings, pro-attitudes toward those meanings, and positing a history of causal relations to a publicly accessible world. To be radically and fundamentally confused concerning one's ascriptions of linguistic behavior to others would entail that a person is confused concerning what constitutes a language or pro-attitude for herself. Because, pace Wittgenstein, there are no private languages, fundamental confusion on the individual level implies fundamental confusion on the communal level since language is a social phenomenon; however, to assert this is to deny the prospects for community, and thereby to contradict an obvious state of affairs of the world around us. In the end, Davidson concludes the 'heady and exotic doctrine' of conceptual relativism cannot be maintained.

Unrecognized by scholars is that Popper in his paper the 'Myth of the Framework'[71] argues a position very similar to Davidson's; however, Popper's approach differs, because although he shares Davidson's concerns about language, his emphasis is fundamentally methodological. The myth of the framework asserts 'a rational and fruitful discussion is impossible unless the participants share a common framework of basic assumptions or, at least, unless they have agreed on such a framework for the purpose of discussion'.[72] Those individuals who accept the myth take it to be 'a logical principle or based on a logical principle'.[73] However, Popper contends that the myth actually undermines rational inquiry. It has its origins in three errors concerning the nature of language and rational discussion. First, it presupposes agreement as the aim among participants in a discussion. Second, it assumes that a greater likelihood of agreement leads to an increase in shared content. Finally, the myth of the framework invites 'standards of mutual understanding which are unrealistically high'[74] and because they cannot be met, the result is cultural relativism and consequently an irrationalism that undermines critical inquiry. Although I'll treat these points in greater detail in Chapter 4 below where I address Popper's views on society and politics, briefly his argument is as follows: To specify a shared framework as either a background condition for discussion or as the aim of discussion is to invite the possibility of violent and totalitarian measures to secure it; if we intend agreement for the sake of agreement as either a means or an end, who is to gainsay how it is attained? Moreover, if the ideal of a shared framework is deemed unattainable, then people may give up on discussion as a way to resolve conflicts. Popper writes, 'I think ... that it [the myth of the framework] is not only a false statement but also a vicious statement which, if widely believed, must undermine the unity of mankind, and so greatly increase the likelihood of violence and of war.'[75] The second error the myth commits is based upon a misunderstanding concerning the relationship between probability and content. Its advocates invoke the myth because it is believed to contribute to the likelihood of an increased understanding between people and thus for greater shared content between them. But Popper argues that the more improbable agreement is, the more our theories say about the world. For Popper this follows because of the inverse relationship between probability and content: as we increase the probability of a shared framework between theories the less the theories say about the world. Finally, the myth of the framework contributes to cultural relativism and eventually irrationalism because it advocates standards for agreement that are unrealistically high, that is, if participants don't agree on all essential points then there is always some basis for disagreement. And for Popper, in the background of this idealized standard of agreement is the all too common commitment to an idealized form of justification.

Ultimately, the problem of language is the problem of understanding how it contributes to the activity of criticism begun on the biological level. Popper understands science as an outgrowth of the biological method of trial and error elimination. Thus to continue the critical activity began on the biological level, the goal is to identify an understanding of language that can help us to comprehend scientific practice and thus the growth of knowledge. To this end, Popper's evolutionary account of language reveals the possibilities for a unique balance concerning language as representation and language as use.

3.4. The Political Problem

For some political theorists, political theory concerns issues of leadership within social units as it affects the stability, scope, and direction of social organization. Thus for some social and political theorists, politics seems to demand that they address the question of the end, aim, or purpose of society; to lead is to lead others somewhere. Moreover, if there is a purpose to history, then those leaders who can unite societal aims with history's purpose are the persons who should rule. Discerning the purpose of history in a manner that facilitates the creation of social policy is aided by principles of historical explanation called historical laws. Historical laws provide a blueprint for the ideal society and help to identify the means for its realization. The best political blueprints are immune to revision and can be immediately implemented.

For Popper, the above approach to political theory leads to the nightmare of totalitarianism; it is to invite a closed society opposed to freedom and human dignity. In *The Open Society and Its Enemies*, Popper's classic treatise on sociology and politics, he argues as follows: The growth of human knowledge precludes the possibility of historical laws because the growth of knowledge changes the historical landscape in a way that defies prediction; however, in closed societies law-likeness is identified with the standards and ideals of the tribe and neither the tribe nor its worldview are to be challenged if social cohesion is to be maintained. Consequently, closed societies try to arrest change by implementing an ideal and static conception of the state that will presumably realize a social utopia when the ideal is actualized in history. Confidence concerning the form of a particular social utopia is supported by the belief that the blueprints for it conform to historical laws, and it is the idea that there are historical laws to guide the selection from competing blueprints that Popper calls historicism. Now a utopian society is most directly and best implemented when the persons in power can start with a clean slate and when they have total and uncensored control of societal organization. Thus individual aims must be sacrificed to the needs of the collective; this includes purging society of individuals and institutions that challenge or do not fit the utopian blueprint and enacting authoritarian measures to secure the ideal at all levels of society. Closed societies make political aims, political practices, and political procedures fixed and mutually supporting such that the imposed and artificial union between the three undermines the distance necessary for one to function as a critical check on the others. In contrast, an open society doesn't impose a union between aims, practices, and procedures; it especially allows for personal decisions by its members because it doesn't fear the novelty that such decisions introduce into society.

Therefore the problem of politics is how to regulate the interrelationship of political aims, political practices, and political procedures in a manner that supports social cohesion while at the same time allowing for a system of rational constraints on how societal organization is realized. Yet in the case of both totalitarian and democratic societies the ideal is for the three to be one: totalitarian aims are attained by totalitarian means and procedures; likewise, in a democracy democratic aims are to be secured by democratic processes. But if the identification between aims, practices, and procedures is to be rejected in the first instance, why is it not to be similarly rejected in the second scenario? Consequently, the challenge for the political scientist is to identify a theory of society sufficient to establish under what conditions the union of aims, practices, and procedures is appropriate and when it is not. An initial response may be to posit that the answer to the above question is located in the domain of ethics, which is prior to and transcends social existence because society is made up of individuals who bring their respective values together to form some type of political and social unity. But Popper rejects such an approach and asserts the autonomy of sociology. Indeed, he quotes Marx approvingly, ' "[i]t is not the consciousness of man that determines his existence – rather, it is his social existence that determines his consciousness" '.[76] Popper's commitment to Marx's epigram is twofold. First, evolution indicates our pre-human ancestors were 'social prior to being human',[77] thus many principles regulating social organization or 'sociological laws' are present in a biological sense as background conditions for human society. Second, Popper rejects psychologism in the manner of Mill's notion of social life as 'the outcome of motives springing from the minds of individual men'[78] because its emphasis on a psychological method makes the issue of the *origin* of personal values and commitments fundamental, and thus leads to historicism. Finally, to reject psychologism and a psychological method as the proper approach to sociology and political theory is not to dismiss the individual. For Popper, political theory must make explicit how personal existence and collective unity can be reconciled. Consistent with his overall method of conjectures and refutations, Popper insists we must formulate 'a political *demand*, or more precisely, a *proposal* for the adoption of a certain policy'.[79] Thus political society or the state is a response to a conjecture, and conjectures do not have to prove their pedigree in terms of an origin or specify an essence before meriting serious consideration. However, on Popper's reckoning, those political theories having the most pernicious influence on twentieth-century life are exactly those theories that have developed historicist philosophies as a pedigree to justify their social policies. Because historicist philosophy and the closed societies that are a consequence of it have been advocated by such prominent Western philosophers as Plato, Hegel, and Marx, exposing the intellectual shortcomings of the historicist's worldview is a task to be undertaken in earnest.

Working with the dichotomy of closed and open societies understood as ideal types,[80] Popper's solution to the problem of politics begins by asking a new question that places the issue of objectivity at its center. Closed societies arrest change in the name of securing an unalterable social framework whose aim, practices, and procedures are immune to public scrutiny. The agents of arrested social development are those individuals who should rule because they best comprehend or represent the archetype that society is seeking to implement. Popper's political theory replaces the question of who should rule with the question of how society can best be structured to replace incompetent rulers. The aim of getting rid of incompetent rulers is best supplemented if utopian social engineering is jettisoned in the name of piecemeal social engineering that is characterized by advancing trial solutions to social problems that are of a restricted scope and do not mandate the overhaul of society. In this way, the tenure of politicians cannot be secured on the grounds that they must remain in power to see a utopian agenda to fruition, because no such agenda is advanced, and thus individuals laying claim to special insights (seers or prophets of destiny) or special pedigrees (bluebloods or aristocrats) are no longer needed. For Popper, an objective political order, that is, one not restricted to the idiosyncratic worldviews of utopian despots, demands the publicity of social policy and a *tradition* of social criticism grounded in the intersubjective testing of public initiatives. The conjecture of limited trial solutions and the demand for publicity and criticizability that function as constraints on such solutions are one with freedom. Together and in a mutually supportive way freedom, publicity, criticizability, and the restricted scope of political solutions facilitate a distance between political aims, practices, and procedures that allow for an objective political order. The result is a public space in which civilization can flourish; however, there is a risk involved because the identity between aims, practices, and procedures characteristic of closed societies is threatened by the 'strain of civilization'[81] that leads people to forsake the open society in the name of the comfort of not having to make decisions for themselves.

3.5. The Ethical Problem

Recent literature on Popper's thought treats of his views on ethics. The focus is long overdue despite the problematic and inconclusive character of Popper's moral claims. Popper's writings indicate a passion and a commitment to put right what he takes to be the intellectual and moral wrongs of his day that can be best identified as an ethical zeal. He unequivocally advocates for persons the duty to be optimistic about improving the human condition, especially along Enlightenment ideals; for persons to work to alleviate human suffering in a manner consistent with the desires of those who are suffering and thus avoid the petty imperialism of mandating what is best for others; and for academics not to shirk the responsibility to be clear and to make relevant to the societies where the privilege of academic life exists, the intellectual problems addressed in their scholarly researches. In his limited writings on ethics Popper argues for both the autonomy of ethics and a negative utilitarianism. On my reckoning, the inconsistency of these two approaches is both glaring and shows Popper's failure to apply the full resources of his philosophical system to the problems of ethics. Why Popper failed to see the difficulties of reconciling two such disparate ethical views is less than completely clear. His emphasis on the autonomy of ethics is one with his belief that a moral agent is radically responsible for the choices and actions she undertakes. However, this rejection of heteronomy along Kantian lines is undermined by a negative utilitarianism that explicitly mandates that the choices an individual makes be reconciled to a cost/benefit analysis in terms of the present and conceivable future status of her own and others material, intellectual, and social well-being. Moreover, it isn't that Popper is per se opposed to meta-ethical inquiry, but he seems less than confident concerning how to reconcile such inquiry with the taking of a decision that for Popper is the focus of a moral life.

My aim is to rectify these shortcomings in Popper's treatment of ethics by use of his overall account of objectivity, and his theory of rationality in particular. Accordingly, I argue that only Popper's thought possesses the resources to answer the fundamental problem of ethics, 'Why be moral?' This problem concerns the relationship between reason and morality, and, unfortunately for the prospects of an adequate solution, traditionally has been answered along justificationist lines. Consequently, I assert that a non-justificationist answer consistent with Popper's critical rationalism is the only proper remedy to the problem of ethics.

Notes

1. Russell (1912).
2. Wittgenstein (1961) 151.
3. Wittgenstein (1968) 51.
4. Ibid.
5. Ibid. 50.
6. Malcolm (1986). This expression is from the title of Malcolm's text.
7. Wittgenstein (1968) 51.
8. Ibid. 50.
9. Kuhn (1977) 121.
10. Rorty (1982).
11. Ibid. xl.
12. Rorty (1989) xv.
13. A central topic for deconstruction is the notion of khora or place. As I understand it, the notion of khora concerns the problem of both delimiting and retaining a context for inquiry. See Derrida and Caputo (1997) Chapter 3.
14. Meyer (2001) 5.
15. See Derrida (1982). Derrida proposes to tympanize philosophy. Here 'tympanize' is equivalent to 'ridicule'.
16. Popper (1959a) 16.
17. Ibid. 15.
18. Popper (1963) 64.
19. Miller (1994) Chapter 3 and (2006) Chapter 6.
20. Hume (1975) 26.
21. Russell (1985) 101, cited in Miller (1994) 1.
22. Popper (1959a) 95.
23. Popper (1959a) and Miller (1994) Chapters 1, 3 and 4.
24. Popper (1959a) 22.
25. Miller (1994) 54.
26. Popper et al. (1974) 31.
27. Popper (1945b) Chapter 12. Popper understands Hegel's absolutism and theory of identity to be a serious threat to critical inquiry and freedom. Hegel's philosophy is an object of harsh criticism for what Popper understands to be its service of a political agenda.
28. Miller (2006) 176.
29. Miller (1994) 2–6.
30. Notturno (2000) 202.

31. Chesterton (1957) 42.
32. Bartley (1984) 114.
33. Ibid. 115.
34. Ibid. 115–16.
35. Popper (1945b) 231.
36. Bartley (1984) 119.
37. For a contrary view see Artigas (1999).
38. Bartley (1984) 238ff.
39. For the important difference between the role of deduction in the growth of our awareness of the truth versus an increase in truth content, see Miller (2006) 69.
40. For the debate over CCR see Miller (1994) Chapter 4.
41. Miller deserves much credit for developing this distinction. See Miller (2006) 72.
42. Miller (1994).
43. Popper (1979) Chapter 9.
44. Miller (1985) 10.
45. Miller (1994).
46. Miller (1974) and Tichy (1974).
47. Gillies (1993) 210 and Sober (1993) 48.
48. Miller summarizes their proof in his (1994) 209–210. For our purposes the key aspect of Miller's summary is paraphrased as follows. If f is false, then the proposition x and the biconditional f-iff-x have opposite truth values. Let f be a false consequence of the false theory H. Allow $Ct(H)$ and $Cf(H)$ to denote the truth content and falsity content of the theory H respectively. If x is part of the $Ct(H)$, then f-iff-x is an element of the $Cf(H)$. As well, the biconditional f-iff-(f-iff-x) is logically equivalent to x. Thus each element in $Ct(H)$ can be correlated with an element in $Cf(H)$.
49. Popper (1959a) 92.
50. Ibid.
51. Ibid. 13. Throughout Popper's writings he emphasizes the importance of the problem situation to the project of criticism. Also, see Miller (1994) 86. Ultimately, our problems have their origin in biology. See Popper (1979) Chapter 1.
52. Popper (1959a) 39.
53. Agassi (1995) 11.
54. Popper (1979) 146ff.
55. Popper (1945a) Chapter 10.
56. Popper (1979) 149.
57. Ibid. 25.
58. Popper (1945b) 238.
59. Popper (1963) 27.
60. Popper (1959a) 15.
61. Popper (1983) 20–27.
62. Popper erroneously attributes this view to Aristotle. See Popper (2001) 3.
63. Popper et al (1974) 977.
64. Popper rejects nominalism as a metaphysical position adequate for science. See Popper (1963) 262–3.

65. Miller (1994) 10–11.
66. See the editor's introduction to Popper (1982b).
67. Platts (1997) 1.
68. Rorty (1999) 269.
69. Popper (1979) 119–20.
70. Davidson (1984b).
71. Popper (1997) 33–64.
72. Ibid. 34–5.
73. Ibid.
74. Ibid. 33.
75. Ibid. 35.
76. Popper (1945b) 89.
77. Ibid. 93.
78. Ibid. 92–3.
79. Ibid. 112.
80. Jarvie and Pralong(1999a).
81. Popper (1945a) 5.

Chapter 2

Scientific Method and Objectivity

Introduction

For Popper, a proper understanding of scientific method is at the heart of his epistemological enterprise and is the mainstay of his response to the epistemological problem of objectivity explained in Chapter 1 above, i.e., how to account for the growth of knowledge. To this end, all philosophy, all science, addresses 'the problem of cosmology: *the problem of understanding the world – including ourselves, and our knowledge, as part of the world*'.[1] Accordingly, to impose order on the complexities of the world around us, science deals in idealizations and oversimplifications.[2] For example, it is well known that Newton's first law of motion specifies a state of affairs that is never actualized in the physical universe, since it asserts that a body at rest or in motion stays at rest or in motion unless acted upon by an external force; however, a state of affairs immune to external forces is never encountered in Newton's world of macroscopic objects. And it is such experience-transcending features of natural science that leads it to be conflated with the claims of metaphysics and thereby invites philosophers to consider the problem of the relationship between the two.[3] Because for many philosophers metaphysical claims were found to impede scientific practice; because scientific training seems able to produce productive natural scientists without first producing in them a mind bent on metaphysical subtleties; because natural science is a democratic undertaking with its own system of checks and balances, it no longer needs a queen of the sciences to bow down before.

Consequently, many scientists and scientifically minded philosophers have sought to eliminate all metaphysical elements from the domain of scientific inquiry. Of particular importance to Popper's philosophical outlook are the logical positivists of the early and middle twentieth century who waged war on metaphysics with crusading zeal to realize what they hoped would be a genuinely scientific worldview. Philosophy is never strictly abandoned by the positivists, but serves to undermine itself in the sense that it would ultimately yield its field of inquiry to scientific speculation informed by an exact scientific method.[4] Clear identification of this method is thereby essential to the positivists' agenda. It is on the point of method as it relates to the problem of demarcating science from metaphysics and pseudo-scientific disciplines that the positivists spurred Popper's own thought in reaction to their Weltanschauung. The positivists' account of scientific method places induction at its core and they use it as a principle to demarcate science from metaphysics. For Popper, this approach won't do. To their credit, the positivists correctly raise the question of the logical relationship between theory and experience. Admittedly, the theoretical claims of the natural sciences must be related to and constrained by experience in some way, but by asserting induction as the heart of the scientific method to distinguish empirical inquiry from metaphysics, the logical positivists simultaneously undermine a central component of science. Popper states:

> It may appear as if the positivists, by drawing this line of demarcation, had succeeded in annihilating metaphysics more completely than older anti-metaphysicists. However, it is not only metaphysics which is annihilated by these methods, but natural science as well. For the laws of nature are no more reducible to observation statements than metaphysical utterances. (Remember the problem of induction!) ... Thus this attempt to draw a line of demarcation collapses.[5]

Accordingly, Popper endeavors to identify a new solution to the problem of demarcation. This undertaking is the wellspring of his reflections on science which establish him as the foremost philosopher of science in the twentieth century. However, we shouldn't lose sight of the fact that Popper's philosophy of science is a particular expression of his desire to understand the more general problem of the relationship between critical thinking and dogmatic thinking that is essential to the growth of knowledge and thus fundamental to the core problematic of epistemology, as he understands it. For Popper, knowledge has a biological basis understood as the attempt by organisms to impose order and regularity on their experience, and in the case of humans, whether this demand for order is satisfied in a manner commensurate with the Inquisition or whether a program emphasizing objectivity and the growth of knowledge will emerge is in no way settled. Popper affirms science to be one of humanity's best hopes for the latter.

In the remainder of the chapter, I examine the two central issues around which Popper develops his account of scientific method: the problem of demarcation and the problem of induction. Moreover, I indicate how a unique instance of the problem of induction (the metaphysical problem) can serve to unify the elements of Popper's thought in a manner that increases our understanding of his treatment of objectivity.

1. Popper's Solution to the Problem of Demarcation

1.1. The General Logical Context

Popper labels the problem of demarcation 'Kant's problem', because it is central to Kant's Enlightenment project of establishing the limits of reason. Like Kant, Popper takes the defense of the rationality of science to be central to the growth of knowledge. Additionally, Popper believes the problem of demarcation to be 'the source of nearly all the other problems of the theory of knowledge'.[6] Specifically, the problem of demarcation leads to a critical examination of the problem of induction (especially the metaphysical problem), which in turn raises the problem of the testability of scientific theories, and this issue entails the problem of demarcating 'rational theories and irrational beliefs'.[7] Now whether Popper is entirely correct concerning the centrality of the problem of demarcation to the theory of knowledge, there can be little doubt that it is essential to his own approach to philosophy. Interestingly, Popper has much to say about the logic of demarcating science from metaphysics and pseudo-science, but given the importance of the topic to his epistemology, it is somewhat surprising that he never identifies the general logical context that gives rise to issues of demarcation. What I mean is, when labeling a state of affairs as scientific in opposition to metaphysical or pseudo-scientific, are we dealing with a dilemma, a simple disjunctive syllogism, or say a decision of the kind studied in rational choice theory? Although Popper consistently reminds his readers that he is not interested in any 'sharp demarcation between science and metaphysics'[8] since scientific and metaphysical insights have aided one another's speculative development; nevertheless, reflecting on the general logical situation in which the problem of demarcation takes shape will only serve to sharpen our appreciation of Popper's solution and the relevance of the problem of demarcation to his treatment of objectivity. Of the three options, a dilemma best reflects the centrality of criticism to the criterion of demarcation Popper identifies.[9] If the general context of demarcation takes the form of a standard constructive dilemma, then central to analyzing it is the locus of the disjunctive propositions. (In the subsequent example reference is made to Popper's notion of falsifiability, to be explained in greater detail in this section. For the present discussion understanding the notion of falsifiability is not essential to the logical point at issue.) Consider the following constructive dilemma:

> If a system of statements belongs to the natural sciences, then it is empirically falsifiable.

If a system of statements is metaphysical, then it is not empirically falsifiable.
Either a system of statements belongs to the natural sciences or it is metaphysical.
Therefore, a system of statements is either empirically falsifiable or not empirically falsifiable.

In this example the conclusion is a tautology, and because any premises or no premises are sufficient to entail a tautology the problem of demarcation would show itself to be trivial if it has the above form. If the positions of the disjunctive statements are switched, then a tautology is still present but now it functions as a premise:

If a system of statements belongs to the natural sciences, then it is empirically falsifiable.
If a system of statements is metaphysical, then it is not empirically falsifiable.
Either a system of statements is empirically falsifiable or it is not empirically falsifiable.
Therefore, a system of statements either belongs to the natural sciences or it is metaphysical.

Because a tautology is a necessarily true statement and can't be proven to be false, the normal procedure when the disjunctive premise of a dilemma is tautologous is to escape the dilemma by criticizing one of its conditional statements; this is known as 'grasping a horn of the dilemma'. Grasping the horn of a dilemma in a logically valid manner takes the form of *modus tollens*, and Popper asserts it is this minimal concession to classical logic that is essential to his deductive method of testing. Moreover, analysis of the role of the disjunctive premise reveals how the adjudication between competing claims can easily get bogged down in issues that preclude criticism from being carried out, and thereby both limits the resources for demarcating one state of affairs from another and, undermines the growth of knowledge. For instance, the disjunctive premise of a dilemma most readily takes the form of a necessarily true statement when one disjunct is the term complement of the other (e.g., $E \vee \sim E$). However, can it be determined independently of critical methods whether a disjunct deserves to be designated as a term complement or in some more informative manner that allows for novel insights and critical thinking? Consider a new dilemma:

If a Democrat gets elected then taxes go up.
If a Republican gets elected then services are cut.

> Either a Democrat or Republican is elected.
> Therefore, either taxes go up or services are cut.

The disjunctive premise avoids becoming a tautology only if we critically engage what it means for a candidate to be, say Republican, in a manner that allows us to assert of her more than 'She's no Democrat' (∼D). Obviously to do this, a person needs to do more than simply label another position in terms of the standards built into her own political worldview. Merely to label one thing in different ways doesn't show it to be different, as Frege's famous example of the Morning star and Evening star demonstrates. Moreover, any attempt to list the differences between Democrats and Republicans seems to require a technique for discerning that an item on one list is not covertly reducible to or synonymous with an item on the other list. Now given Popper's rejection of induction there is no way to justify such differences by empirical means. Most importantly, some philosophies make it impossible to move the disjunctive premise beyond the uninformative and vacuous qualities characteristic of a tautology. For example, a philosophy that advocates meaning-as-use renders arbitrary the question of whether one disjunct deserves to be labeled a term complement (∼D) or designated in a content-rich and therefore more criticizable way as some other position (R). For the only resource such a philosophy has available is to report or describe linguistic usage, and if a person's level of use stops at the claim 'She's no Democrat', then a philosophy based on meaning analysis can't explain why inquiry should proceed in a way that contributes to our learning more about what it is to be a Republican. As was shown above, criticism is still possible even when the disjunctive premise is a tautology, but a commitment to a genuinely informative demarcation between science and metaphysics in the name of the growth of knowledge shouldn't preclude the identification of as much criticizable content as possible.

1.2. Empirical Refutability

Consequently, the form of a dilemma highlights the merits of a critical methodology, and in distinguishing science from metaphysics and pseudo-science Popper maintains genuinely scientific claims are subject to deductive testing by experience, that is, they are open to empirical refutation.[10] In this way, experience itself 'appears as a distinctive method whereby one theoretical system may be distinguished from others; so that empirical science seems to be characterized not only by its logical form, but, in addition, by its distinctive *method*'.[11] Whereas the positivists sought to relate theory and experience by means of logically invalid inductive inferences intended to explain how generalizations can be built up from singular statements, Popper argues experience constrains our theoretical claims in a deductive, yet negative manner:

> In other words: I shall not require of a scientific system that it shall be capable of being singled out, once and for all, in a positive sense; but I shall require that its logical form shall be such that it can be singled out, by means of empirical tests, in a negative sense: *it must be possible for an empirical scientific system to be refuted by experience.*[12]

Thus, Popper's criterion of demarcation demands of the natural sciences falsifiability by empirical means. Popper identifies two distinct senses of the expression 'falsifiability'.[13] In one sense, the falsifiability of a theory denotes a logical relation between a theory and its class of potential falsifiers, or basic statements. Basic statements *'have the form of singular existential statements'*.[14] This means that basic statements designate a determinate space/time region where an observable event can occur. Key to Popper's deductive theory of testing is that basic statements can be derived from a universal theory and a set of initial conditions; moreover, in the case of a falsifiable theory its class of potential falsifiers is non-empty, that is, there exist basic statements that can contradict the theory in question. Thus falsifiability in the first sense refers to requirements for theory construction; a scientific theory must be constructed so to be open to possible empirical refutation. To this end, our conjectures about the physical world must exhibit both boldness and avoid any kind of 'conventionalist stratagem'.[15] A bold theory says as much about the world as possible, and thus it has greater testability since the more a theory references the world, the more states of affairs there are to count possibly against it. Popper came to see that the relationship between boldness, informative content, and testability could be expressed in reference to the probability calculus. He asserts an inverse relation between probability and content. Let Ct = informative content and P = a probability measurement consistent with an axiomatization of the probability calculus, then:

$$Cta \leq Ctab \geq Ctb$$

$$Pa \geq Pab \leq Pb\,[16]$$

Therefore, the more improbable or bold a theory is the greater its content, and thus the more testable it is. As well, Popper's demand for bold and novel insights in part motivates his rejection of essentialism. Essentialism can be used to immunize a theory from empirical refutation. For example, if a person asserts that all swans are white, and if her generalization is confronted with the refractory evidence of a black swan, then she can always assert that to be a swan is to *be* white and thus reject any state of affairs that is inconsistent with her account of what constitutes the essence of a swan.[17] Consequently, novel developments and new areas of inquiry are dismissed.

Such stultifying features are characteristic of other conventionalist stratagems Popper decries. In general, conventionalism is an approach to science that treats science as a system of statements taking the form of 'implicit definitions'.[18] Natural laws do not capture the simplicity present in the world but are the simplest way logically to construct a world amenable to scientific measurement. 'According to the conventionalist point of view, laws of nature are not falsifiable by observation; for they are needed to determine what an observation and, more especially, what a scientific measurement is.'[19] Basically, conventionalism is a straightforward apriorism, and like all apriorisms it eschews empirical constraints. Needless to say, because the subject at issue is *empirical* science this is problematic at a fundamental level. And the same is true of other particular conventionalist stratagems such as explaining away criticisms by accusing one's critics of 'inadequate mastery of the system.... Or ... by suggesting *ad hoc* the adoption of certain auxiliary hypotheses, or perhaps of certain corrections to our measuring instruments.'[20] As previously acknowledged in Chapter 1 above, Popper admits the presence of a priori elements in science especially at the level of theory construction. Indeed, for Popper, theories are not inferred from experience in any way, instead they are bold guesses about the cosmos and our place in it. And if the theoretical claims of empirical science are to retain their ties to experience and not merely address 'an infinite number – of "logically possible worlds"' then our guesses must be regulated by the '"world of our experience"'.[21] Ultimately, Popper's account of scientific method can be summed up as asserting science to be a system of controlled guesswork, where the controls are to be found through the successful merger of both logic and experience.[22] If we look at the failure of the inductivist approach to scientific method, it correctly gives experience, in the sense of testable singular statements, a role in science, yet because an inductivist account of science is committed to a justificationist agenda it does so at the expense of inviting an infinite regress and so unsettles our logical sensibilities. In contrast, Popper's account of scientific method properly sets its sights on the role experience must play in scientific methodology. He asks us to consider an important distinction between an account of science where theories are inter-subjectively testable ad infinitum and an account of science that invites an infinite regress of justification.[23] In an inductivist account of science an infinite regress occurs, and experience, so to speak, falls out of sight, because each statement advanced as evidence requires the support of some prior testable statement and so on to infinity. Popper's account of science, however, allows for testing ad infinitum. If a basic statement is identified as a falsifying statement for a universal theory, then devoid of any problem situation that would give inquirers a reason to challenge the falsifying basic statement's scientific merit, the universal theory and its attendant initial conditions are as a unit accepted as falsified.[24] However, nothing about his deductive method of testing precludes the issue of the scientific status of either the theory or the falsifying basic statement from being taken up again. Conclusions in science are tentative, but this doesn't mean they are arbitrary. The decision tentatively to accept a falsification in the first place, the recognition of the problem situation that led inquirers to challenge the falsification, and the decision to renew inquiry in light of a new problem situation are each advanced only if they allow for an increase in the testability of the scientific claims at issue. Thus although tests may continue ad infinitum, the decision to continue testing is not arbitrary because the demand for increased testability is constrained by experience as reflected in both the aim of learning more about the world of our experience and because it is *empirical* science that is practiced. In this way experience is never removed from the decision to test or not to test a theory, and because strict proof or disproof is not at issue an infinite regress is not possible. Thus, Popper asserts that if his use of the term falsifiable is taken in a second sense to denote conclusive falsification of a theory, he never intended it to be understood in such a manner.[25]

a. Strictures on Scientific Testing

a.1. Intersubjectivity and Repeatability

Scientific tests are qualified in two interrelated and important ways. A theory is to be constructed so to allow for intersubjective testing, that is, 'in principle it can be tested and understood by anybody'.[26] Moreover, intersubjective testing mandates that test are repeatable, because without this characteristic any ' "occult effect" ... one for whose reproduction he [the scientist] could give no instructions',[27] could be admitted to science in a way that they couldn't be extirpated and to do so blurs the line between subjective conviction (I know I saw that!) and objectivity. Here, intersubjectivity is not a mere amassing of opinions, nor is it simply to ask others, 'Did you see what I saw?', although this comes closer to the mark. Intersubjectivity must not be understood simply as a means of getting others involved in the appraisal of theories, since this could be done by coercion, bribery, sex appeal, or other sociological factors. Rather, intersubjectivity is the demand for methodological constraints so that subjective convictions, or factors based upon individual nuances such as wearing eyeglasses, or singular elements such as an unexpected power surge affecting test apparatuses, can be subject to controls that allow for their elimination. In empirical science these controls take the form of repeatable experiments. Historically, intersubjectivity, subjectivity, and truth have been locked in a strange dance that has permitted issues of origins to set the tempo. The idea is that if all subjective, singular, and therefore arbitrary elements are removed from scientific testing, then the result will be an account of the world as logically, epistemically, and causally independent of how a knowing agent cognizes it. Thus getting rid of such elements will show a theory to be based in the world itself, and because the aim of empirical science is to represent the world truthfully, then such an account will be devoid of error. But Popper was the first to see that evolutionary theory breaks step with the idea that it is the origin of a theory or basic statement that makes it true, that is, true if its origin is the world and false if it has its origin within the subject. For what Popper was among the first to understand is that evolutionary theory teaches the student of scientific method that human adaptations understood as theories or expectations about the environment don't always fit with the way the world is; yet, such adaptations *are provided by the world* via our evolutionary history as organisms. Thus it's not the origin of a theory that determines its objectivity in the sense that there is a fundamental cleavage between the mind-independent (objective) and the mind-dependent (subjective) because the very way in which we frame our perspective on the world is a function of the framework the world itself provides us with.[28] Accordingly, Popper emphasizes the intersubjective testability of theories and a correspondence theory of truth as key to his account of objectivity. A judgment of correspondence in science is a meta-level judgment that isn't determined by asking the source of our knowledge claims, but follows upon deductive testing in reference to an experimental set-up. As stressed earlier in Chapter 1 above, justification and truth are distinct, so the source of an inquirer's knowledge is neither a criterion for truth nor a vindication for her knowledge claims.

a.2. Theory-ladenness and Conceivability

It's important to recognize that from his earliest writings Popper acknowledges both the theory-laden aspect of observational statements, and as a human activity scientific methodology involves decisions and value judgments on the part of its practitioners. Unfortunately, advocates of the strong program in the sociology of knowledge have exploited these very limited nods to human agency to deny the objectivity of science.[29] For such thinkers the procedure has been to demand immodest goals in the form of both trans-historical standards of justification and truth holism, while concomitantly asserting the historically conditioned and therefore very modest perspective that is available to a person or community at a given point in time. By the expression 'truth holism' I mean the view that a statement is true if and only if we can see all the various ways it is true, that is, truth is of the whole. Truth holism allows advocates of the sociology of knowledge to take an unhealthy approach to the role of revisionism as part of rational inquiry. There is a difference between revisionism as a methodological tool to aid classification and revisionism as a tool to forestall the classification of statements as true. The importance of this distinction can be highlighted if we consider the role thought experiments play in the scientist's revision of scientific claims in contrast to the way the sociologist of knowledge uses them to argue that truth is arbitrary. The role of thought experiments in both science and history trades on how they respectively address the relationship between conceivability and possibility as part of making revisions. However, let me say at the outset that although I think this line of inquiry is important to follow because arguments based on the relationship between conceivability and possibility are not infrequently exploited for improper ends; nevertheless, as a general principle the claim that conceivability implies possibility is erroneous. But that it is erroneous does not affect the scientific approach to revisions at all, while the situation is less sanguine for advocates of the sociology of knowledge.[30]

Sociologists of knowledge in appealing to the historically conditioned nature of sociological claims may argue, for example, that particular consequences of the French Revolution are not to be classified as true because it is conceivable and therefore possible for a subsequent understanding of say, economics, to force a revision of present assessments of those consequences. Therefore, on such reckoning a sociologist of knowledge forgoes talk of true statements, because the whole truth is not known. This approach to the role of revisions in rational inquiry both incorporates truth holism and takes a consideration of mere alternatives grounded in the presumed relationship between conceivability and possibility to suffice as genuine critique. In contrast, science doesn't treat as serviceable for scientific practice open-ended appeals to what is possible; instead, the classification of statements as true or false requires that possibilities be constrained by what is intersubjectively testable, and once they are identified, natural laws serve as constraints on what is admitted as scientifically possible.[31] Confusion sets in concerning the difference between the historically conditioned claims advanced by the sociologist of knowledge and the demand for testability in the empirical sciences, because actualizing the demand for empirical testability is often contingent upon historically contingent developments. For example, the ability to test a particular theory can be dependent upon as yet unrealized technological innovations that are deemed possible because conceivable, but not actual at the present time. However, the historical dimension of testability in empirical science here at issue, that is, how things are testable in a conceivable and therefore possible future, differs greatly from the sociologist's use of history, especially as she uses the relationship between conceivability and possibility to forgo the classification of statements as true or false. In empirical science claims about what is possible are constrained by natural laws; thus, if a singular event can't be subsumed under a universal statement, 'i.e., hypotheses of the character of natural laws'[32] (often used as background knowledge), then it isn't possible to test that singular statement. Thus, while a theory can still be classified as true or false we do so in a dogmatic or uncritical fashion, and while admitting statements to science in such a way is not at all irrational, since acceptance of conjectures is not a matter of pedigree, still this is only one half of the scientific process because the process is only complete when a conjecture is submitted to rigorous empirical testing, and this testing requires that empirical observation statements be subsumed under a natural law.

In contrast, there are no historical laws.[33] Accordingly, in the absence of historical laws the conceivability-entails-possibility thesis functions for the sociologist of knowledge as a fallback position, that is, as the only accessible constraint on the prognostication of a future singular event, and in turn the claims about the conceivable future inform her approach to revisionism. Thus together with truth holism the conceivability-entails-possibility thesis for the sociologist of knowledge serves to forestall the classification of statements as true or false. With some charity, we can say that this approach is praiseworthy to the extent it is motivated by the recognition of the fallibility of our truth claims. Nevertheless, because there are no historical laws history unfolds in an unpredictable manner, and the fallibilism asserted by the sociology of knowledge is characterized by a fundamental complacency or wait and see attitude that is foreign to scientific inquiry. But as Miller points out, 'sitting around complacently with a well-meant resolve to accept any refutations that happen to arise is a caricature of genuine falsificationism'.[34] Simply put, the conceivability-entails-possibility thesis exposes critical inquiry to armchair pedantry if it is not used in conjunction with a system of constraints such as the natural laws of empirical science. This lack of constraint is heightened when it is further supported by claims alleging the indeterminacy of translation or other problems of meaning.

Surveying the argument advanced thus far in this chapter, we can now see that Popper offers a methodological solution to the epistemological problems he raises. Empirical science is distinct from metaphysics, pseudoscience, etc., by its method. This method requires the statements of science to be treated not as indicative claims but as guesses or conjectures about the world. These conjectures must be advanced in a manner that exposes them to be discounted because of empirical testing. Theory acceptance is, well, just that, theory acceptance. The scientific researcher accepts a theory and then does her best to disconfirm it. Miller captures in a very insightful way how it is that Popper's methodology revolutionizes the problem of empirical scientific knowledge. The traditional program of empiricism has been to account for how it is that experience is productive of knowledge. Simply put, Miller sums up this aspect of the empiricist program as: 'WYLIWYK *what you learn is what you know*.'[35] However, Hume's critique of induction reveals that, especially in the case of our most informative knowledge claims, the universal generalizations of natural science, this quite simply isn't the case, since what we learn from experience is never sufficient to establish the scientific generalizations that make up the more interesting part of scientific knowledge. The converse of the above is WYKIWYL *what you know is what you learn*. And Miller argues that the truly revolutionary aspect of Popper's thought is that Popper rejects this element of empiricism. 'Turning his back on traditional empiricism, Popper decisively separated the categories of knowledge and learning.'[36] For Popper, we have to know something before we learn anything.[37] Thus, we advance conjectures as something we presume to know and then expose what we claim to know to empirical refutation. Although, Miller doesn't go into any detail explaining the a priori nature of our conjectures I'll turn to this subject in the next chapter where I explore aspects of Popper's evolutionary epistemology and the biological origins of knowledge. Miller's above analysis is an extension of the following point made in his 1994:

> We might put things this way: when a theory fails a test, we learn something but end up knowing nothing (since what we knew, our theory, has been eliminated). But when a theory passes a test (when, that is to say, it is corroborated), we learn nothing (since we already knew what the result of the test was going to be) but we continue to know something.[38]

In working out how it is that science enables us to learn more about the world, Popper's methodology first and foremost identifies and explores the logic of empirical science, since the relationship between theory and experience presupposes a formal structure that demands clarification, hence the title of his classic treatise on scientific method, *The Logic of Scientific Discovery*. However, that his approach to scientific method emphasizes the importance of logic does not mean that he is unaware of or uninterested in the sociological and normative dimensions of scientific activity. Popper writes, '[a]s I tried to make clear in 1934 (L. Sc. D., p. 37; and sections 10 and 11), I do not regard methodology as an empirical discipline, to be tested, perhaps, by the facts of the history of science. It is, rather, a philosophical – a metaphysical – discipline, perhaps partly even a normative proposal.'[39] Accordingly, in what follows treatment of particular issues central to the problem of demarcation will help to indicate how it is that the normative elements of science arise from within science and are not merely imposed from without. This point is important given the extent to which many philosophers relativize science to antecedent societal values. One thinker who provides a sustained treatment of the sociological dimension Popper's thought is Jarvie. Jarvie, in an original and challenging monograph, *The Republic of Science: The Emergence of Popper's Social View of Science 1935–1945*, argues that Popper's account of science stands in need of a 'sociological demarcation'.[40] He grounds this conclusion in Popper's appeal to various methodological rules as conventions requisite to overcome the shortcomings of a purely logical demarcation between science, pseudo-science and pre-scientific theories (myth, folklore etc.).[41] However, Jarvie's analysis is supplemented by several striking claims that on my reckoning indicate the limitations of his sociological approach to Popper's thought. Jarvie contends that Popper came to realize sociology is 'transcendentally necessary' to an account of scientific rationality.[42] This couldn't be further from the truth if 'transcendentally' is understood to denote a valid condition for the possibility of a given experience, in this case empirical science. Popper rejects any a priori valid conditions for science or rational inquiry in general, (and what else could 'transcendentally necessary' mean?) since such conditions would have to be beyond criticism. Furthermore, in an attempt to harmonize his own views concerning the sociological tenets of Popper's thought and what he takes to be one feature of Popper's objectivism, Jarvie writes: 'Scientific ideas and results are articulated in language, and those statements are subject to the laws of logic.'[43] However, Popper dismisses the idea that there are laws of logic since this would entail essentialism.[44] Jarvie's central claim is 'that thinking socially is central to Popper's philosophical enterprise'.[45] In itself this assertion is mostly uncontroversial, but Jarvie identifies social with institutional and grounds this view in the fact that scientists share methodological procedures and decisions. Thus he seems to think that this sharing or intersubjectivity invites '*an institutional approach to the philosophy of science*'[46] but on my reckoning it is an adventurous leap from intersubjective to institutional. Consider the case of the Curies' isolation of pure radium. Marie and Pierre Curie were two scientists who in their shared research exercised the critical rationality characteristic of genuine scientific practice. They were rigorously critical of one another's work, but it would be difficult to characterize their isolation of pure radium as an institutional undertaking. Concerning her parents' segregation from institutional structures Eve Curie writes:

> Out of the traditions and principles of the French Revolution, which had created the metric system, founded the Normal School, and encouraged science in many circumstances, the State seemed to have retained, after more than a century, only the deplorable words pronounced by Fouquier-Tinville at the trial in which Lavoisier was condemned to the guillotine: "The Republic has no need for scientists".[47]

The point of this example concerning the Curies' isolation of radium is that while science, as Popper argues, cannot be a Crusonian undertaking, the intersubjectivity characteristic of scientific method need not be an institutional arrangement, and this seems to be especially true when groundbreaking researches are at issue. Most importantly, Jarvie's appeal to institutional structures itself needs a criterion of demarcation. How much and how widespread must a sharing of aims and methodological procedures be to call an arrangement genuinely institutional and genuinely scientific? Jarvie correctly identifies the analogy Popper asserts between science and the game of chess in *Logic*, page 53, and concerning Popper's use of this analogy he writes:

> Certainly he [Popper] seems to be arguing that science is constituted by its rules, as is chess. He also seems to be allowing that the rules of science can be debated, hence they are not immutable. Much the same goes for chess. The rules of chess have evolved and might evolve more. A rule revision would not necessarily make for a new game, especially if the rule were adopted by the International Chess Federation.[48]

Now the rules of chess and chess strategy are logical undertakings that have received institutional support through the ICF, but the Federation's support doesn't make any particular strategy or alteration of the rules of chess more or less rational; nor does an institutional approach to chess serve to demarcate it from other board games in a way that wasn't sufficiently determined by the logic of chess (not in the sense of pure logic) prior to the Federation's appearance. An occasion for chess playing provided by the ICF is not a cause of chess playing; only its system of rules and conjectured strategies are; so too, institutional structures are an occasion for science, not the real thing. Moreover, Jarvie's final sentence in the above quoted paragraph has an authoritarian ring quite foreign to Popper's philosophy. In fact, in the opening pages of *The Logic of Scientific Discovery*, Popper's rejoinder to Reichenbach's assertion that the 'whole of science' accepts induction is 'after all, "the whole of science" might err', and induction is to be rejected because it leads 'to logical inconsistencies'.[49] Thus it is logic and experience and not the issue of whether a theory of inference has been adopted by institutional structures (i.e., 'the whole of science') that should guide conclusions concerning scientific methodology.

1.3. Consideration of Criticisms

In the remainder of this section I want to examine some particular issues central to Popper's account of demarcation insofar as they touch upon the problem of objectivity. The first three issues: (1) the problem of asymmetry, (2) the problem of the empirical basis, and (3) the problem of the repeatability of tests are well-rehearsed aspects of Popper's thought treated by various thinkers familiar with his work. Still, a consideration of these topics is worthy of another go around if only because they are the first stumbling blocks to a deeper appreciation of Popper's epistemology. The last problem I want to address in this section examines objectivity and demarcation as indebted to the notion of intersubjective testing. Recently, Saul Kripke's work has been interpreted as containing a serious objection to the claim that objectivity can be grounded in intersubjective tests.

a. Asymmetry

In *The Logic of Scientific Discovery*, Popper argues that there is a fundamental asymmetry between falsifiability and verifiability.[50] As criteria of demarcation, each asserts a logically different relation between singular statements and universal statements. Given the logical form of universal statements, such statements can never be established (i.e., verified) as true by a finite set of true singular statements; however, a true singular statement can contradict a universal statement and thus 'it is possible by means of a purely deductive inference ... to argue from the truth of singular statements to the falsity of universal statements'.[51] However, a genuine asymmetry is often denied for the following reason.[52] It is an undisputed point of deductive logic that the falsity of a given statement entails the truth of its contradictory opposite. Thus if a universal statement is falsified, the pure existential statement that contradicts it is thereby verified. Therefore, the falsification of every universal theory requires a corresponding verification because the falsifying basic statement must itself be true if the universal statement is false, and so every falsification contains an implicit verification. Consequently, falsifiability can't stand apart from verifiability as a genuine criterion of demarcation to distinguish metaphysics and pseudo-science from science. Moreover, its failure as a criterion of demarcation seems to be further supported by the fact that the pure existential statement that is identified as verified has the same scope as the claims of metaphysics.[53] An unrestricted existential statement, such as, 'There exists a 10lb diamond', treats of all of reality; since a 10lb diamond may have existed in the past, it may currently exist, or it could exist in the future. Thus a pure existential statement that is the contradictory of a falsified universal statement both functions like a claim of metaphysics and seems to be verified whenever a universal statement is falsified. The end result is Popper's demarcation criterion between metaphysics and empirical science seems to achieve nothing that it is intended to accomplish.

In examining Popper's response to the above criticism we must begin by acknowledging the following terminological point. For the purposes of this discussion the words 'falsification' and 'falsifiability' refer to a purely logical relationship between a theory and its class of basic statements.[54] Now unfortunately, Popper's critics are confused concerning what it is that logic alone is able to provide concerning scientific practice. What isn't properly appreciated is that although a falsification is contradicted by a verification this fact references how statements are to be compared and subsequently classified. Here the classification of a pure existential statement as true and hence 'verified', and the classification of the universal statement as false, is a mere description of the logical situation. In no way does the description of the logical situation *establish* a pure existential statement as true. Inductive procedures – that is, inferences based on a repetition of experience or an accumulation of test statements – are nowhere to be found; yet this is exactly what the verification of a statement requires. Most importantly, pure existential statements are not testable independent of a conjectured universal theory and a set of initial conditions. Considered in isolation from these other elements required for theory construction, pure existential statements are not unlike 'occult effects' and are insufficient to specify testable conditions, including testable conditions that could lead to their verification, if such a state of affairs were possible. Additionally, Popper argues singular existential expressions such as 'Here is a glass of water', contain universal concepts and hence evidence a law-like disposition, what Popper calls 'the transcendence inherent in any description'.[55]

Scientific Method and Objectivity   49

Thus examined both from the perspective of their relationship to the other statements that renders them testable (i.e., universal statements and initial conditions) and from the perspective of their internal makeup, pure existential statements reveal their dependence upon statements having a universal form if they are to be at all testable. Thus, it is only because of their deductive relationship to statements having a universal form that pure existential statements are transformed into the basic statements that make them adequate for scientific testing and thus part of scientific methodology. To sum up, singular existential statements independent of their relationship to universal generalizations are inadequate to the task of scientific testing; once it is understood that scientific methodology investigates the truth of universal generalizations, it then becomes obvious that pure existential statements are insufficient to establish the universal statements that enable such singular statements to have a home in scientific practice; consequently, singular statements are logically relevant to universal generalizations only to the extent singular statements identify a class of potential falsifiers. The classification of scientific statements is all-important to scientific practice, but if empirical science is meant to tell us about the world, then classification independent of empirical scientific testing quite simply isn't sufficient to characterize genuine science. Accordingly, a mere description of the logical relationship between the statements of science is only helpful if such a description facilitates an understanding of how we classify scientific statements in light of empirical tests, and the only methodology we can avail ourselves of if singular statements are to have a valid logical relationship to the universal generalizations that are the object of scientific inquiry, and thus have a role in science, is a critical methodology. The end result is the asymmetry Popper asserts to be the case.

b. Empirical Basis

Popper emphasizes singular statements as potential falsifiers of scientific theories to indicate the empirical character of such theories, but to do so raises the question of the empirical character of singular statements themselves. He writes: 'We have now reduced the question of the falsifiability of theories to that of the falsifiability of those singular statements which I have called basic statements. But what kind of singular statements are these basic statements? How can they be falsified?'[56]

Popper labels this problem the problem of the empirical basis, and his response again takes up the relationship between singular statements and universal generalizations. Keeping in mind the above discussion of asymmetry, Popper's position is that singular statements can be divided into two types: (1) pure existential statements and (2) basic statements. Basic statements, as noted above, are statements derivable from a universal generalization and a set of initial conditions. As well, basic statements specify a determinate space-time region where an observable event can occur. The general question now becomes in what sense are basic statements representative of the empirical character of science, in plain words, how do they show science to have a basis in experience. In *The Logic of Scientific Discovery*, Popper structures his treatment of the problem of the empirical basis around Fries' trilemma. Fries asserts that the problem of the empirical basis forces one to choose between an infinite regress of proofs, dogmatism, or psychologism.[57] Popper contends Fries and most other epistemologists opt for psychologism because it provides scientific claims with a foundation in sense experience. Thus, Popper begins his discussion of the empirical basis of science with a critique of psychologism in what we might call its naive and sophisticated forms. Naive psychologism is the view that the empirical basis of science is to be traced to the perceptual experiences of the scientist. Thus on such an account, '[s]cience is merely an attempt to classify and describe this perceptual knowledge, these immediate experiences whose truth we cannot doubt; *it is the systematic presentation of our immediate convictions*'.[58] Popper concludes this approach founders on the problem of induction, since scientific generalizations can't be inferred from our immediate convictions even if they are true, and naive psychologism is unable to account for the role universal expressions play as part of descriptive statements, since the use of universal expressions 'cannot be correlated with any specific sense-experience'.[59] Continuing his critique, Popper argues that the theory of protocol sentences developed by Neurath and Carnap is simply a sophisticated form of psychologism. They were correct to assert that science doesn't proceed by the direct comparison of statements with reality and thus advanced protocol sentences as records of one's immediate observations. Moreover, Neurath posits that protocols are not sacrosanct and can be subject to revision. However, this last point 'leads nowhere if it is not followed up by another step: we need a set of rules to limit the arbitrariness of "deleting" (or else "accepting") a protocol sentence. Neurath fails to give any such rules and thus unwittingly throws empiricism overboard.'[60] Not without irony, Popper's own approach to the problem of the empirical basis is confronted with the similar charge of taking the empirical out of empirical science.

Thus, having rejected psychologism, what exactly is Popper's solution to the problem of the empirical basis? Properly understood, Popper's solution simultaneously adopts variants of all three elements of Fries' trilemma. Basic statements 'have admittedly the character of dogmas'[61] because although not an unrestricted fiat they are nevertheless a decree giving expression to the outcome of the decision-making processes of practicing scientists; however, because the acceptance of a basic statement is not intended to prove anything, testing may proceed ad infinitum without inviting an infinite regress of justification, thus acceptance of a basic statement is not doctrinaire acceptance; also, basic statements may be causally linked to a knower's perceptual experiences 'but a basic statement cannot be justified by them [i.e., the perceptual experiences] – no more than by thumping the table'.[62]

On my reckoning, although there is much that is salutary in the account of basic statements developed in *The Logic of Scientific Discovery*, Popper's analysis isn't complete until conjoined with his propensity theory of probability and causality developed in his later writings.[63] Popper posits observability as a material requirement for basic statements.[64] At first glance, this requirement seems to reintroduce psychologism into the procedure of scientific testing since observations seem to imply an observer. However, Popper argues that in the same way observations by one sense can in principle be translated into tests involving another sense, so too the notion of an observable event can be replaced or taken to be synonymous with, for example, 'an event involving position and movement of macroscopic physical bodies'.[65] Therefore, he concludes '[o]bservations and perceptions may be psychological, but observability is not'.[66] What Popper has in mind is that very often technological devices (e.g., Geiger-counters, seismographs) can and have replaced the role of human perceptual experiences in registering the outcome of events or as aids to predictability.[67] Most importantly, in relation to his later theory of propensities, observability like a propensity can be understood to be a function of the physical situation, specifically the experimental set-up. The relationship between Popper's account of the empirical basis and his theory of propensities can be expressed as follows: Observability : Empirical Basis :: Propensities : Probability/Causality. Here, I simply want to emphasize that Popper's account of observability in *The Logic of Scientific Discovery* is amenable to interpretation as a physical propensity, and although the details of his account of propensities is provided in the next chapter, the central point for now is that it is the physical situation with its testable conditions and not the scientist(s) as observer that is primary.

Popper's treatment of the interrelationship between the objective features of scientific statements and the decisions scientific researchers make concerning such statements is often confused, as it is by Keuth in his recent monograph on Popper's thought. Utilizing texts he believes highlight the conventional and therefore subjective aspect of Popper's treatment of the empirical basis, Keuth quotes Popper as follows concerning where a consensus by scientists is to lead us:

> at a procedure according to which we stop only at a kind of statement that is especially easy to test. For it means that we are stopping at statements about whose acceptance or rejection the various investigators are likely to reach agreement.... If this too [a continuation or repetition of empirical tests] leads to no result, then we might say that the statements in question were not intersubjectively testable, or that we were *not*, after all, dealing with *observable events*.[68]

Keuth then goes on to comment, '[b]ut why should events be nonobservable if their observation does not lead to unanimous judgment; why should a statement not be intersubjectively testable if there is no agreement on the result of its test?'[69] But this isn't Popper's point at all. Agreement doesn't determine observability. Observability is a function of the ease with which scientific researchers can test a statement, and this ease is determined by the empirical content and the logical form of the statement at issue (i.e., self-consistent, non-tautologous). Keuth's confusion on the role of free decisions in science is highlighted when he writes, 'judgments about observation statements are *free decisions* only insofar as they cannot, from the logical point of view, be justified'.[70] But why does Keuth assert this? Obviously a person is free to reject a justified state of affairs no less than an unjustified one. Keuth thinks otherwise not because he doubts the freedom of scientists to make decisions; rather, he thinks there is something about the content of a justification that necessarily motivates the decisions scientists take concerning it. But the same is true for Popper concerning the role of intersubjectively testable basic statements in his account of science. Claims concerning observability are part of the procedure of testing which takes as its aim a particular problem situation and the empirical content and logical form of the statement under consideration. The decision to continue testing or to repeat a test does involve a decision, but it's not merely by taking a decision that the matter is concluded. Insofar as what is at issue is empirical science, the content of statements at issue must have the final say.

Like his general approach to epistemology, Popper's account of basic statements is intended to avoid both foundationalism and arbitrariness. At first glance, he doesn't seem to help his case against the arbitrariness of the empirical basis when he asserts the relativity of basic statements.[71] But upon closer inspection, Popper's point is that the empirical basis is not relative to the knower in the sense that his/her judgment or perspective decides the matter, rather the decision to accept a test or to stop testing is made relative to the stage of development in the critical discussion. In the case of science, critical discussion as well as scientific objectivity are concomitant with intersubjective testing. Thus the commitment to intersubjectively test a theory isn't per se an antecedent value, but a standard that arises as part of scientific practice. The demand for intersubjective testing occurs with our recognition of our own fallibility. That we are less than pleased with the results of our perceptual and cognitive judgments and apparatuses invites us to check the method employed when asserting an answer to a problem. This check must increase the testability of the theory at issue. Although emphasis is given to how intersubjectivity arises as part of scientific practice, what is not claimed here is anything paralleling the justificationist tendency to make induction apiece with rationality and thereby settling the issues of scientific method and demarcation by fiat. Miller writes, '[t]owards a better methodology of science science itself will lead the way'.[72] This is the simple truth. The tendency on the part of philosophers to identify some prior commitment to inform the adoption of one methodology or one theory over another is part of the pernicious justificationist mindset that culminates in foundationalism. The problem situation that invites the development of a particular methodology is little different than an itch that begs to be scratched. Problems like itches are irritants, they get under the skin, but addressing an itch is not indicative of an antecedent commitment to realism, empiricism or the correspondence theory of truth. These commitments are the result of meta-level judgments that are the consequence of reflecting on actual practice, a subliming of what is at hand. And an account of scientific method is such a meta-level activity. We like to think hindsight is 20/20, that our meta-level judgments reveal implicit values or aspects of scientific practice that are assumed or imposed at the beginning. (So-and-so is obviously thinking of or valuing X even though she is unaware of it!) Concerning literary and philosophical works the deconstructionists, to their own minds anyways, have made a virtue of seeing what a particular author fails to see in her own work. Every text contains the seeds of its own undoing in what is not being said, that is, in the implicit halves of the dualities that make the text possible. Accordingly, the deconstructionist mindset is something like, 'Father, forgive them, they know not what they do.' The end result is a prescience that can impugn a work regardless of the problem situation it attempts to address. Science, however, makes use of basic statements to establish a preference between competing theories and in this way, if the basic statement is to be relevant to adjudicating between the competing theories, the problem situation can't itself waver or else there is no competition. Thus, once theories are brought into competition with one another the question of values is systematically constrained by the act of inquiry informed by the problem situation and by the available resources amenable to empirical inquiry.

c. Repeatability

The project of empirical refutation demands the repeatability of tests as part of intersubjective testing. Only if a scientific statement is subject to repeatable tests is it genuinely falsifiable. And although the repeatability of empirical tests is invoked as part of Popper's solution to the problem of demarcation, on a prima facie level it looks as if the topic of repeatability forces one to consider that other central problem of scientific method, the problem of induction. The repeatability of tests appears to mandate the adoption of the principle that nature continues uniformly the same, and because such a claim asserts that the future will continue to be like the past it seems overtly inductive. Hesse, for example, writes:

> Again, one past falsification of a generalization does not imply that the generalization is false in *future* instances. To assume that it will be falsified in similar circumstances is to make an inductive assumption, and without this assumption there is no reason why we should not continue to rely upon all falsified generalizations.[73]

Popper's response is that 'repeatability entails independence of predecessors',[74] thus no inductive inference is at work here in the sense that a future repetition of a falsifying test is inferred from past occurrences. And although Popper advances the above claim concerning the independence of repetitions in the context of discussing probability statements, his point applies to scientific statements in general because given his propensity interpretation of probability to be explored in the next chapter, causal statements on the propensity interpretation can be interpreted as probabilistic statements having the probability of 1. As well, arguing from the perspective of Popper's general methodology, Miller asserts a rejoinder to the repeatability problem that quite rightly plays up the conjectural element of Popper's thought. A generalization that a present falsification will hold in the future can be advanced as a guess subject to refutation to the extent it is rendered testable.[75] Keuth addresses the role of the uniformity of nature in Popper's theory of corroboration and asserts, 'Popper's tacit assumption of the uniformity of nature does not imply that his theory of corroboration involves a principle of induction, for this assumption is not suited to be such a principle.'[76] However, I think there is a logical point concerning empirical testing that needs to be taken into account. Testable conditions result because a prediction or retrodiction is the consequence of a universal theory and a set of initial conditions. Because the content of the statements involved is empirical content, philosophers of science have understood the requirement of repeatability to reference an aspect of experience, viz., the uniformity of nature and not a feature of the logical form of the statements at issue. However, what is at issue concerning the repeatability of tests is the logical relationship between a statement and its expansion. Falsifying basic statements have the logical form of existential statements. Consider an existential statement of the form $(\exists x)(Sx \wedge Px)$ and given a three-member universe of discourse or domain (a, b, c) the existential statement's expansion is $(Sa \wedge Pa) \vee (Sb \wedge Pb) \vee (Sc \wedge Pc)$. Logically, a statement's expansion is neither assumed nor inferred, instead it is described, and in an expansion of an existential statement each element expresses the same logical form as the existential statement itself (i.e., a conjunction). Now, in the same way that we don't expect an individual element of an expansion to have a different form than its parent statement, so too, for logical reasons, insofar as the repetition of a falsification of a theory is an instance of the initial falsifying statement, we don't expect any statement in its universe of discourse to differ. Thus it isn't that we assume or infer nature to be uniformly the same in future instances; rather, subsequent falsifying instances are uniformly the same because they are instances of the initial falsifying statement's universe of discourse. We may of course be wrong in how we identify membership in an expansion, but ascriptions of membership in an expansion can themselves always be tested. However, as we've already seen, to do so is to proceed in a deductive and not in an inductive manner.

d. *Kripke and Intersubjectivity*

Recently, Saul Kripke's work on Wittgenstein's treatment of rule following has been interpreted as denying the relevance of intersubjectivity to objectivity. Deluty asserts:

> In the *Philosophical Investigations* para. 242, Wiitgenstein asserts paradoxically that objectivity is not lost even though communication requires the interplay of agreement in definitions and agreement in judgments. Although Wittgenstein does claim that objectivity is determined only by this interplay, the objective status of logic initially appears to have disappeared. Wittgenstein here foresees the criticism launched by Kripke that objectivity has been replaced by inter-subjectivity.[77]

Given the prominence of Kripke in analytic philosophical circles Deluty's charge is important. However, without getting sidetracked by the details of

Deluty's presentation which I disagree with at key points concerning its interpretation of Kripke, I want to explain why Kripke is entitled to be read as denying the relationship between intersubjectivity and objectivity, and to advance a response along Popperian lines.

In his book *Wittgenstein on Rules and Private Language*, Kripke can be read as both agreeing with Wittgenstein's rejection of truth conditions as a basis for objectivity while at the same time disagreeing with Wittgenstein's alternative assertibility conditions that are apiece with his account of meaning as use. The end result is that none of the options explored in the text provide a basis for objectivity and the idea that objectivity is to be secured by agreement between subjects is deemed untenable.

Kripke's position develops as part of his discussion of Wittgenstein's treatment of rules. He states his aim is neither straightforward exegesis, nor to advance his own critical evaluation of the subject matter at issue, but instead to present 'Wittgenstein's argument as it struck Kripke, as it presented a problem for him'.[78] Accordingly, the text is something like a personal report aimed at producing 'an elementary exposition';[79] however Kripke at times wavers between advocate and reluctant critic.[80] Obviously, Kripke doesn't intend the relevance of his analysis to the philosophical community to be based on the fact that it conveys *his* opinion; rather, he argues for two original claims: (1) Wittgenstein's private language argument occurs much earlier in the *Philosophical Investigations* than scholars have traditionally asserted, and (2) the private language argument is a response to a skeptical paradox, 'a new form of scepticism ... the most radical and original sceptical problem that philosophy has seen to date'.[81] It is in working out how the private language argument is a response to the skeptical paradox that Kripke can be read as denying the relevance of intersubjectivity to objectivity. Wittgenstein, in characteristically cryptic style, states the paradox as follows: 'this was our paradox: no course of action could be determined by a rule, because every course of action can be made out to accord with the rule'.[82] Throughout his exposition Kripke formulates the paradox in reference to the following problem situation. Is it possible to fix the rules governing use of the addition symbol so that when computing the addition problem 68 + 57 equals some number, we get 125 because 'addition' means plus, or we get 5 because addition specifies 'quus'? At the heart of the skeptical paradox is the question whether there exists a 'superlative fact' that can serve as a rule of interpretation to identify that a speaker ascribes one meaning to an expression over another. Kripke explains that Wittgenstein uses the private language argument to establish that no occurrent fact or interior disposition cognizable by introspection could suffice to legitimize one meaning over another. Most pertinent to Popper's philosophy of science, however, is Kripke's account of why there exist no conditions of an external character adequate to serve as a superlative meaning-fact to adjudicate between competing meanings for a given expression. Concerning external constraints, Wittgenstein, according to Kripke, doesn't advance a skepticism having a form such that a person would engage in 'sceptical denials of our ordinary assertions'.[83] To do so would make it impossible to get the paradox off the ground, since 'for the sceptic to converse with me at all, we must have a common language'.[84] Thus the everyday public fact of language is not at issue; instead, the issue is how is it possible to identify truth conditions for a rule of interpretation such that people can ascribe meaning to terms in a manner that makes language at all possible.[85] Kripke contends that Wittgenstein's private language argument establishes that the pursuit of truth conditions must be forsaken if the skeptical paradox is to be avoided. This follows because as mentioned above there is no fact ascertainable by introspection that could serve to establish one meaning over another. Furthermore, since we do not want to say that the truth conditions for an expression are to be determined by the public fact of intersubjective agreement, the pursuit of truth conditions must be replaced.

Wittgenstein replaces the aim of specifying truth conditions for a statement with the aim of identifying assertibility conditions for an expression. Assertibility conditions express the circumstances under which members of a language game are justified in *using* certain expressions. Thus:

> What follows from these assertibility conditions is *not* that the answer everyone gives to an addition problem is, by definition, the correct one, but rather the platitude that, if everyone agrees upon a certain answer, then no one will feel justified in calling the answer wrong.[86]

Moreover, issues of usage can be checked in reference to societal practices but if this tempts one to think objectivity is to be correlated with truth conditions sufficient to secure a superlative meaning-fact, then Kripke concludes that Wittgenstein contends such an agenda leads us back to the skeptical paradox.

When considering the importance of Kripke's treatment of Wittgenstein for Popper's theory of scientific method, it is important in the name of fairness to acknowledge that Kripke doesn't understand the issue to be an epistemological one. He thinks Wittgenstein is not raising a skeptical problem concerning how one *knows* that there is a meaning-fact, but whether there is such a fact to be accounted for in terms of truth conditions. In this way, the problem is structured so as to have an objective character, yet the conclusion is that objectivity can never be secured by any intersubjectively approachable fact.

The assertibility conditions that are part of the solution to the skeptical paradox emphasize intersubjective agreement and 'checkability – on one person's ability to test whether another uses a term as he does'.[87] Hence, it seems that the only cause for disagreement concerning the merits of intersubjectivity is that it is no longer directed at truth conditions but takes issues of meaning in the sense of meaning-as-use and thereby assertibility conditions as paramount. But this is not the case.

Kripke argues against Wittgenstein's position in the following way. Given the rejection of truth conditions, a test whether two persons agree in language usage is not in any way a measure of truth or falsity, thus at best it can underwrite a report about agreement. When all is said and done, the conclusion Kripke's Wittgenstein arrives at is the following: agreement in use satisfying assertibility conditions equals language and language equals agreement-in-use satisfying assertibility conditions. Leaving to one side the glaring tautology that no one seems to acknowledge and sticking to the topic of rule formulation, the following analysis results. A rule that tests for agreement among assertibility conditions cannot be prior to language because it is formulated in language; however, this means the rule satisfies some standard for assertibility conditions. Additionally, any rule for assertibility conditions would have to specify standards by which agreement obtains and when it doesn't obtain, but this means that besides containing assertibility conditions a language would have to possess rules *about* its assertibility conditions. But as Kripke points out, this requirement 'falls foul of Wittgenstein's strictures on "a rule for interpreting a rule"'.[88] In the end, neither the pursuit of a superlative meaning-fact able to be the object of intersubjective tests nor assertibility condition that can function as an objective constraint are adequate for objectivity; and most importantly, subjective agreement without either seems arbitrary.

There are many avenues of criticism available to Popper given Kripke's approach to both Wittgenstein's problem situation and solution. Kripke's exposition shows the problem situation and its solution to be structured in terms of issues of foundationism and justificationism. Neither of these issues is unrelated to Popper's account of scientific method, and Popperians could well fight these thinkers concerning the proper role of truth conditions to problem solving, philosophical and otherwise; however, the focus here is on intersubjectivity and I propose that Kripke's inquiry is guided by an aim that accounts for why he fails to understand properly the relationship between intersubjectivity and objectivity. What both Kripke and Kripke's Wittgenstein fail to grasp is that there is a fundamental difference between testing carried out such that it can promote a report and testing performed in the name of advancing criticizability. As philosophers of language, both Kripke and Wittgenstein take as their aim to illuminate how language functions, and this ultimately terminates in reports about user practices. Indeed, Kripke and Kripke's Wittgenstein both inquire into a superlative fact specifying truth conditions and assertibility conditions specifying agreement in use to promote a report about how users of a language may test for the appropriateness of one meaning for an expression over another. In contrast to Popper, inquiry terminates in a report on language use that serves as an explanation of user practices and that is the end of it, whereas, for Popper, inquiry also aims at an explanation; however, more importantly, inquiry and the explanations that result aim at an increase in testability. The value of intersubjectivity as providing objectivity is that it can promote an increase in testability showing such an increase to occur in ways that are not reducible to subjective agreement. Before explaining Popper's approach to increased testability and intersubjectivity let me add that I think Kripke rejects both truth conditions and assertibility conditions as adequate to promote a report about testing, but the key point for Popperians is that we don't want a report, we want increased testability, and this can only be had via intersubjectivity.

Concerning the empirical testing of hypotheses, Popper identifies something he labels the 'law of diminishing returns'[89]. The idea here is that in conducting scientific tests inquirers want tests that are severe rather than lax; this means such tests will be risky, that is, the tests will aim at addressing as much empirical content as possible. To construct risky tests inquirers have to take into account their background knowledge, but this also means that the results of new tests are incorporated into that background knowledge, and as our background knowledge increases our tests become less and less severe, thus ' "the empirical content of a theory becomes stale after some time" '.[90] However, background knowledge is a springboard to empirical tests and it best ceases to be stale when the content it provides us with is itself the conclusion of rigorous intersubjective testing. Consequently, our background knowledge serves to increase the testability of our theories. Thus for Popper, background knowledge is a constraint on hypotheses and the tests pursued in investigating them.

Test for agreement-in-use is conducted relative to background knowledge, but independent of intersubjective testing there is nothing to prohibit that tests conducted in reference to such knowledge represent anything more than the eccentricities of the person conducting the test. For Kripke's Wittgenstein, agreement-in-use facilitates a report concerning whether or not a person, say Jones, satisfies the assertibility conditions for a language or, as Wittgenstein sometimes expresses it, a form of life. However, it is difficult to see how a check for agreement that denies the relevance of intersubjectivity can in any meaningful way serve as a severe and therefore genuinely informative and objective test.

When a person checks Jones' linguistic activity for agreement in use she does so in relation to her own background knowledge or linguistic framework. Such a check can generate a report that can begin to form the interlocutor's background knowledge about Jones, but because the background knowledge is simply a reflection of the interlocutor's creativity or lack thereof in asking questions, it is completely arbitrary. Given that the interlocutor starts with a theory concerning how best to question, there is no way of exhausting its content or to attain 'staleness' because there is no constraint concerning the originality and relevance of a question without background knowledge made possible by intersubjectively testable constraints. Additionally, without such constraints it is impossible to gauge whether testing is severe. Intersubjectivity provides objective grounds for any tests that can be called a test. Consistent with Kripke's own approach I wish to stress that what is not at issue here is how an interlocutor *knows* her questions are tests, but what could constitute a test without considerations of intersubjectivity regardless of whether truth conditions or assertibility conditions are at issue.

2. The Problem of Induction

2.1. The Solution: Conjectural Knowledge

'Hume's problem simply does not arise for guesses',[91] argues David Miller, summing up the central aspect of Popper's response to the logical problem of induction. Unfortunately, the importance of the conjectural component of Popper's philosophy is too often unappreciated. If the pursuit of empirical evidence is aimed at taking the guesswork out of science by establishing a theory as true it is insufficient to the task; indeed, this point is exactly what Hume's logical critique of induction is meant to highlight. But as already emphasized in Chapter 1 above this doesn't mean scientific conjectures are unconstrained. Scientific generalizations have the form of guesses and may be eradicated by empirical evidence, and this point is exactly what Popper's emphasis on empirical refutations is meant to highlight. Simply put, recognizing that scientific claims are conjectural completely changes our understanding of scientific method. No longer are observation statements pursued in the name of establishing, justifying or providing good reasons for scientific theories. This fact is often misunderstood because there is a tendency to understand guesswork as incorporating some form of inference or guesswork is understood to need support because a person must have reasons for advancing one guess rather than another. Obviously, once a guess is advanced one person can ask another why she advanced this conjecture rather than that, but for a person to fail to have a response doesn't make her guess any less a guess. Indeed, the natural tendency is to label such a state of affairs a guess in its truest form. In the background of the issue of the role of guesswork in science is the problem of theory acceptance. Relative to empirical data, theory acceptance has both an anterior and posterior dimension. On an inductivist account of scientific method a theory is first accepted as worthy of scientific investigation because it is understood to have a basis in experience and is built up from empirical data. For an inductivist, after a theory is accepted as worthy of scientific investigation in light of its experiential groundings, it is subsequently incorporated among the body of scientific truths when enough or sufficient empirical data (however this is to be understood) is seen to support it above any rival theories. Popper's approach to theory acceptance is radically different and is coextensive with his treatment of scientific claims as conjectural. Popper argues 'that a hypothesis can only be empirically *tested* – and only *after* it has been advanced'.[92] For Popper there is no anterior empirical basis for theory construction and once a theory is advanced empirical data is never entertained so as to secure or establish a theory since, pace Hume, this is impossible. However, Musgrave can be understood as arguing that Popper's methodology of conjecture and refutations as a response to Hume is an attempt to have the best of both worlds. Popper agrees with Hume's logical critique of induction; however, Musgrave writes '[y]ou do not answer Hume by agreeing with him'.[93] Because Popper does agree with Hume he asserts that scientific theories are best understood not as indicative claims clamoring for support but as conjectures to be tested. It seems likely that Musgrave would conclude that all Popper has done by answering Hume is to switch the question, and so not really answer Hume at all. Keuth is unambiguous on this point, 'Popper did not *solve* the problem; rather, he *shifted* it.'[94] In a fundamental sense, what is at issue between Popper and his critics is the scope of the question being raised. For Popper, the problem of induction is advanced as a solution to the general problem of scientific method – the logical relationship between theory and experience. Now what Popper argues is that induction is an irrational account of this relationship, but the relationship is not per se irrational, that is, scientific theories or guesses can be rationally constrained in a negative way by empirical tests. Thus what Popper did concerning the general problem of scientific method – the logical relationship between theory and experience – is not shift the problem; instead, he uncovered the true character of universal scientific claims as conjectural. Recognizing the conjectural quality of scientific generalizations requires a different way of understanding the relationship between theory and experience and of scientific rationality itself. Accordingly, it is no more fair to accuse Popper of switching the issue than it is fair to accuse, say, Andrew Wiley of failing to solve Fermat's last theorem because his proof reworks the problem along geometrical lines.

Suffice it to say, Popper's novel understanding of scientific statements as conjectures is sufficient to resolve the problem of induction and Popper's detailed formulations of the various forms the problem of induction can take and his rigorous replies to those formulations is simply to accept the additional challenge of answering inductivists and show that they fall on their own principles. That Popper does this is a boon to his defense of objectivity because it lays bare the objective requirements for the testing of our conjectures.

2.2 The Four Interrelated Problems of Induction

a. The Logical Problem

For Popper the heart of Hume's logical critique of induction is that there is no logically valid way to 'draw *verifying* inferences from observations to theories'.[95] Hume shows that any attempt to establish an inferential relationship between observations and a universal theory leads to an infinite regress and Popper argues that it leads to either an infinite regress or a priorism;[96] consequently, there can be no general criterion for verifying the truth of a scientific theory based upon experience. Indeed, any such criterion would take the form of an inductive principle having a greater scope and therefore be more difficult to establish by observational data than the universal statements of natural science that are to be verified by means of it. Miller states:

> Indeed, if our goal is simply to sort out what is true, the detour through probability or confirmation, or support in the company of the imaginary principle P [a principle of probabilistic/inductive support], is plainly gratuitous. For rather than spend our time trying to classify some absurdly general principle like P as true we would do better to investigate, and try to classify as true (or as false) those much more manageable, yet also much more interesting, factual statements that are at the centre of our concerns, namely genuine scientific hypotheses.[97]

Now although Hume's logical critique argues it is impossible validly to establish universal theories from observations, Popper asserts Hume's criticism leaves 'open the possibility that we may draw falsifying inferences: an inference from the truth of an observation statement ("This is a black swan") to the falsity of a theory ("All swans are white") can be deductively perfectly valid'.[98] Moreover, Popper's falsificationist approach both preserves the principle of empiricism (only experience and observations should determine the truth or falsity of scientific theories) and identifies a way of establishing a preference between theories since we prefer those theories that have withstood rigorous testing versus those that have not withstood such testing; such theories are said to be corroborated.

b. The Epistemological Problem

Concerning the second and the third problems of induction there seems to be some confusion. On page 71 of *Realism and the Aim of Science*, Popper identifies the order of the problems he addresses as logical, methodological, epistemological, and metaphysical. Yet the second problem concerns the problem of rational belief and thus is decidedly epistemological. Popper states:

> The 'problem of rational belief' as the second phase of our problem may be called, is in my opinion less fundamental and interesting than the first. It arises as follows. Even if we admit there is no logical difficulty in showing that how observations may sometimes help us to distinguish between 'good' and 'bad' theories, we must insist that no explanation has been given of the trustworthiness of science, or of the fact that *it is reasonable to believe* in its result – in theories which are well tested by observations.[99]

In *Realism and the Aim of Science* Popper places little emphasis on this phase of the problem of induction for two reasons: (1) he is satisfied that his philosophy of science explains why the pursuit of epistemic states such as belief is inconsistent with scientific methodology, and (2) he erroneously concludes that his theory of verisimilitude is sufficient to render the problem of rational belief interesting in a way it wasn't in the past. However, the argument in *Realism and the Aim of Science* predates Miller and Tichy's independent refutations of Popper's theory of verisimilitude, and so it is more informative to focus on Popper's critique of belief epistemologies.

Popper argues that the problem of rational belief has been the central problem of epistemology since the Reformation.[100] The fundamental problem is how can a person adjudicate between 'competing theories and beliefs', and the standard response is an individual accepts those beliefs that have greater justification than any competing belief.[101] Popper rejects belief epistemologies along with justificationism because both undermine the rationality of theoretical and practical inquiry. Concerning natural science, if we understand both the justification for belief in a universal theory and the content of the theory itself to be underwritten by an inductive inference from experience, then any such belief is irrational given the logical problem of induction. Moreover, if the aim of inquiry is to secure certain beliefs as opposed to doubtful ones, then the project becomes trivial since a person can always remain unshaken in her convictions by refusing to consider opposing evidence and by only counting favorable evidence as scientifically relevant. Popper coined the phrase 'Oedipus effect'[102] to reference the influence a prediction has on an inquirer's approach to the gathering of evidence for the predicted event. In the case of Oedipus the prediction that he would murder his father and marry his mother led him to undertake actions that confirmed that outcome. Analogously, the belief a person has informs how she approaches the evidence before her. Beliefs, like predictions, influence the procedure of gathering evidence; especially when beliefs take the form of convictions there is a natural tendency to look for evidence that confirms one's viewpoint. Thus, Popper's emphasis on scientific statements as guesses is a radical epistemological shift in comparison to traditional belief epistemologies because it informs a whole new approach to inquiry. The recognition that theories are guesses invites a rejection of authoritarianism and foundationalism, since guesses are hardly taken to convey an authoritative standpoint and commonly a person advances a guess when she has no basis for her views. But whereas traditional epistemology emphasizes having a basis for one's beliefs that invites the identification of an authoritative criterion, Popper asserts guesswork is the true form of scientific theories. Popper's most dramatic rejection of belief epistemologies and their attendant psychologism is his World Three theory of autonomous statements-in-themselves. Although his weakest philosophical work is to be found in defense of World Three, it serves to express the important insight that scientists investigate systems of statements independent of whether they have any beliefs in them. Indeed, Miller deserves much credit for emphasizing that the aim of science is the classification of statements as either true or false and not the collection or the establishing of beliefs. Musgrave, however, disagrees. He writes:

> Miller prefers to talk, not of beliefs, but of the acceptance or rejection of hypotheses, and of classifying hypotheses as true or false. As I already said ... nothing is changed by this terminological fad. To accept a hypothesis as true is to believe it (and to accept it tentatively is to believe

Scientific Method and Objectivity 65

it tentatively). To classify a hypothesis as true is also to believe it (and to classify it tentatively is to believe it tentatively).[103]

Musgrave goes on to add that Miller 'sets aside occasional remarks of Popper's ... remarks to the effect that theory preferences can be justified by good reasons, though theories cannot'.[104] His final salvo takes this form: 'He [Miller] complains that I do not divulge what use can be made of claims to the reasonableness of beliefs. Goodness me, the epistemic problem of problems is "What should I believe?"'[105]

To begin, it is difficult to reconcile Musgrave's contentions with Popper's thought. Indeed, Musgrave has to gainsay Popper's World Three ontology of statements-in-themselves.[106] Leaving aside the issue of Popper's particular arguments for World Three, there can be little doubt that Popper advocates the content of beliefs independent of acts of believing as the proper object of the theory of knowledge. Popper consistently points to Einstein's refusal to believe his own theories of special and general relativity and that Einstein's detachment doesn't affect the scientific status of his ideas, because Einstein's belief or lack of belief in them doesn't affect their status as testable conjectures. Moreover, it seems strange in the era of the modern computer, which can both store and be critical of propositions having objective content, that Musgrave concludes consideration of epistemic states is somehow important to the growth of knowledge. For example, the field of evolutionary computation involves the computer simulation of random variation and selection in an iterated process in which conjectured solutions to a problem generate results that are subsequently assessed for fitness in light of a specified goal.[107] Here critical activity is engaged without the presence of belief states. Instead, the computer model progresses toward a solution by placing the objective content of trial answers in competition with one another. Finally, in contrast to what Musgrave says, Miller does address the issue of theory preference on pages 8–9 of his 1994 book by quoting from pages 21–2 of Popper's *Objective Knowledge*, wherein Popper makes no reference to belief states but to the acceptance of the best tested theory as a basis for signaling a preference between competing hypotheses. I agree with Musgrave that Hume's problem of induction does concern the problem of rational belief, but to point out as Popper and Miller have done that Hume takes up an improper focus is not to run from Hume's critique by changing the problem. Musgrave's assertion, '[l]et us be clear about one thing, however; the theory of rationality is, of its essence a theory about the "second world" a psychologistic theory', is simply dogmatic and drags epistemology back to the faculty theories of knowledge and the 'new way of ideas' of the seventeenth and eighteenth centuries that Popper argued 'should be replaced by a more "objective" and a less genetic method'.[108]

c. The Methodological Problem

The third phase of the problem of induction raises the question, ' "How do you know that the future will be like the past?" ',[109] Popper's response, consistent with his rejection of justificationism, is that he doesn't know if knowing involves having a good reason for such an inference. The inference that the future will be like the past concerns the procedure of scientific testing, that is, inductivists assert scientists to make use of this inference in carrying out scientific experiments. More exactly, the methodological issue concerns how a scientist should use the outcomes of past scientific tests when it comes to planning subsequent scientific endeavors. Popper asserts that scientific theories that have survived the most stringent tests devised are said to be corroborated, but corroboration of a theory is nothing more than a report concerning the past performance of the theory at issue. That a theory is highly corroborated is not a guarantee of its future ability to withstand refutation. It is not because they underwrite an inference from the past to the future that scientists use the past test reports of a scientific theory; rather, past test reports are used because science is interested in the truth about the physical world and so scientists prefer those theories that to date have not been classified as false or superseded by a theory of greater explanatory or predictive power.[110] Thus Popper explains how a scientist can, without contradiction, both apply a theory to a given context (say an engineering project) and at the same time not contend that the theory being applied will survive in the future:

> The matter can also be put like this. The question of survival of a theory is a matter pertaining to its historical fate, and thus to the history of science. On the other hand, its use for prediction is a matter connected with its application. These two questions are related, but not intimately. For we often apply theories without any hesitation even if they are dead – that is, falsified – as long as they are sufficiently good approximations for the purpose in hand. Thus there is nothing paradoxical in my readiness to bet on applications of a theory combined with a refusal to bet on the survival of the same theory.[111]

Popper's point is that Newtonian physics has not survived as a true theory of physical space, but nevertheless it's adequate for bridge construction. In the same way that inductivists both acknowledge the importance of singular observation statements to science but are wrong about the role of such statements in scientific method, so too are inductivists wrong about the role of past tests of scientific theories. Such tests play a role in corroboration, a topic singled out for special consideration in the final section of this chapter.

d. *The Metaphysical Problem: The Unifying Aspect of Popper's Theory of Objectivity*

Popper's solution to the problem of induction is his '*deductive method of testing*'.[112] Moreover, Popper's assertion in *The Logic of Scientific Discovery* that his object of inquiry is the logic of knowledge[113] has led both his critics and supporters to emphasize his solution to the logical problem of induction. However, Popper believes that he solves the first three problems of induction thus far examined, but that it is his solution to the metaphysical problem that tends to prevent his solution to the first three problems from being accepted.[114] Unfortunately the metaphysical problem of induction has been largely ignored by Popper scholars,[115] and so their ignorance of it inhibits their understanding of: (1) Popper's overall solution to the problem of induction, (2) the relationship between his early writings on epistemology, his theory of propensities, his later evolutionary epistemology, and most importantly, (3) his treatment of objectivity. To understand Popper's treatment of objectivity one must comprehend the fundamental importance of the metaphysical problem of induction to his account of scientific method. What is the metaphysical problem of induction? Popper expresses it as follows:

> Hume, if not an avowed idealist, was at least a sceptic as to the reality of the physical world. His scepticism was closely connected with his views about induction. He admitted the strength of our belief in a physical world ordered by laws, but asserted that this belief was unfounded. This suggests that the fourth stage of the question should have been: 'I believe that we live in a real world, and in one exhibiting some kind of structural order which presents itself to us in the form of laws. Can you show that this belief is reasonable?' The issue raised here is that of metaphysical realism, in a form which does not so much stress the existence of physical bodies as the existence of laws. For physical bodies are only an aspect of the law-like structure of the world which alone guarantees their (relative) permanence; which means, on the other hand, that the existence of physical bodies (about which Hume was so sceptical) entails that of objective physical regularities.[116]

And again under the label 'Newton's Problem':

> Newton was led, by his theory of action at a distance, to the belief that space was the sensorium of God. The argument is somewhat fantastic, no doubt; but there is more to it than meets the eye. ... Distances in the universe are tremendous. Action at a distance would mean that gravitational effects were, like the Deity, omnipresent in the whole world. ... Einstein solved this problem, or so it seems, by his theory which makes gravitational forces spread with the velocity of light. This solution ... indicates a possible unification of the theories of light and gravity, and it does so by interpreting light, and gravitational disturbances, in terms of structural properties – field properties – of the universe, of our world. And yet, we are still faced with Newton's problem. For what about these structural properties of our world themselves? They are, we believe, the same everywhere and at all times. How are we to understand this?[117]

In a more succinct statement, Popper asserts that the metaphysical problem of induction raises the question of ' "whether true natural laws exist?" '.[118] To say that there is a true natural law is to talk about the universe as such, i.e., its very structure. Because it is a pure existential claim, the issue raised by this question is akin to a metaphysical problem because it asks about something that can't be established by experience. Looked at from the perspective of Humean skepticism, the metaphysical problem asks how it is that physical bodies entail the existence of objective physical laws and at the same time physical bodies presuppose the existence of such laws. Both accounts of the problem indicate that the purported reality of natural laws refers to the added presence of more general structural properties of the physical universe. Popper's choice to use Newton's treatment of space to express what is at issue is most appropriate. For Newton, space serves as an arena where gravitational effects take place; that is, it provides an underlying structural basis to physical events. Analogously, all law-like regularities have a similar structural basis, and any one law does not capture this structure.

At first glance, it is difficult to understand in what sense this problem is important for our treatment of Popper's account of objectivity, or his philosophy in general. Popper states that the metaphysical problem of induction is 'academic' without any 'methodological significance' and its 'solution is not needed to establish' his claim to have solved the first three problems of induction.[119] However, the metaphysical problem of induction is important to Popper's account of objectivity in light of the aim of science, i.e., the classification of scientific statements as true or false.

Popper contends that inductivism commits its proponents to metaphysical views and thus to an account of science which not only prohibits the demarcation of science from non-science, but is incompatible with the aim of getting to the truth about the world. Popper contends the metaphysical implications of Hume's logical problem of induction are grounded in Hume's structuring the problem in terms of a belief epistemology and thus culminate in idealism:

> The empiricist philosopher's belief 'that all knowledge is derived from sense experience' leads with necessity to the view that all knowledge must be knowledge either of our present sense experience (Hume's 'ideas of impressions') or of our past sense experience (Hume's 'ideas of reflection'). Thus all knowledge becomes knowledge of what is in our minds. *On this subjective basis, no objective theory can be built*: the world becomes the totality of my ideas, my dreams.[120]

Although Popper's picture of idealism is that perhaps a little extreme, his point is that Hume's metaphysical views entail experience ceases to function as a corrective to scientific hypotheses. Concomitant with Hume's idealism is his skepticism about physical objects. Now physical objects can be understood as a concatenation of various physical laws, yet Hume argues induction isn't able to establish the truth of even one physical law. However, as already made clear, for Popper the basic statements that make possible scientific tests are identified as such to the extent they are subsumable under universal generalizations having the form of natural laws. This means the very act of scientific testing, which takes physical objects and their properties as the subject matter of investigation, presupposes something about the structural properties of the physical universe, and this is especially true for Popper given his claim that even singular observation statements express law-like dispositional properties. Therefore a solution to the metaphysical problem of induction must reconcile empiricism and metaphysical realism if the scientific method is to carry out the process of critical testing that Popper assigns to it.

e. *Popper's Solution to the Metaphysical Problem*

Initially, Popper believed the metaphysical problem was insoluble, but over time he came to believe that he found in part a solution to it.[121] However, before turning to this topic I want briefly to examine the application of his critical rationalism to metaphysics. Concerning metaphysics Popper advances the radical and nearly paradoxical claim that all metaphysical theories are equally irrefutable but some metaphysical theories deserve to be classified as false.[122] Consistent with his position asserting the tentative nature of all knowledge claims any classification of metaphysical theories will be provisional, however, what is at issue here is how such classification is carried out. Popper develops his thesis concerning the irrefutability of metaphysical claims by analogy to pure existential statements. In general, the irrefutability of statements can be approached from two perspectives: logically irrefutable and empirically irrefutable. The former according to Popper is equivalent to logically consistent, and 'the truth of a theory cannot possibly be inferred from its consistency'.[123] Here it should be emphasized that Popper is referring to the internal or self-consistency of a theory, while the issue of the consistency of metaphysical claims with other theories, what one might call external consistency, is key to his explanation of how metaphysical claims can be classified as false. Empirical irrefutability denotes that a claim about the physical world is 'compatible with every possible experience',[124] that is, no experience can count against it. Pure existential claims when logically coherent are irrefutable in just these two ways, and Popper argues that a metaphysical claim, for example determinism, has the same form: '*There exists* a true description of the present state of this man which would suffice (in conjunction with natural laws) for the prediction of his future actions.'[125] However, despite their irrefutability, metaphysical claims can be critically evaluated and even tentatively classified as false in two ways: (1) in light of the problem situation that gave rise to them, and (2) in reference to other theories accepted as well-corroborated background knowledge. Most importantly, it is in reference to this second point that Popper works out his solution to the metaphysical problem of induction.

Concerning the first point Popper writes:

> Now if we look upon a theory as a proposed solution to a set of problems then the theory immediately lends itself to critical discussion – even if it is non-empirical and irrefutable. For we can now ask such questions as: Does it solve the problem? Does it solve it better than other theories? Has it perhaps merely shifted the problem? Is the solution simple? ... Questions of this kind show the critical discussion even of irrefutable theories may well be possible.[126]

In regards to his second point, Popper argues that metaphysical claims are constrained by those theories that in light of corroboration are classified as true and form part of our background knowledge. For example, Popper asserts Hume's idealism is the result of his reliance on 'a sensualistic theory of knowledge and learning'.[127] And while idealism per se is irrefutable,

Hume's version can be attacked by pointing out that there are better theories of knowledge acquisition than Hume's which posits the reduction of all ideas to sense impressions. As mentioned in Chapter 1 above, Miller uses the expression coat-tailing to explain how metaphysical elements within science are constrained. Miller writes:

> [O]ne may note ... the pervasiveness of metaphysical and other untestable hypotheses within science. Such hypotheses are often introduced as essential adjuncts to scientific hypotheses; indeed, all falsifiable hypotheses have amongst their consequences a host of unfalsifiable statements (ranging from tautologies and unrestricted existential statements to meaty metaphysics) that enter science as it were on the coat-tails of their parents. But these unfalsifiable consequences – to the extent that that is all that they are – are not scientific in their own right; their title is one of courtesy. If their parents are rejected from the realm of scientific knowledge, they will have to be rejected too.[128]

Miller acknowledges as well the extent to which Popper asserts metaphysical insights to influence and contribute to the development of scientific theories. The question of the range of influence of metaphysical insights raises a central issue for Popper's theory of scientific method and its answer takes the form of a methodological decision in light of the aim of science. Popper states, '[w]e can combine the two, empiricism and metaphysical realism, if only we take seriously the hypothetical character of all "scientific knowledge" and the critical character of all rational discussion'.[129] Key here is the idea that empirical science enjoins the scientist to incorporate metaphysical claims into the domain of science to the extent metaphysical commitments advance, or are at least consistent with critical inquiry. Whether legend or not, it was the metaphysical convictions of those professors and clerics at the University of Pisa that led them to refuse to look through Galileo's telescope, since in light of their metaphysical commitments they knew in advance the nature of the solar system. Now the above account of the metaphysical problem of induction indicates that structural questions about the universe are essential to empirical testability. How this is so is made manifest by Popper's solution to the metaphysical problem of determinism. However, it is important to acknowledge at the outset that the following analysis is not an investigation into Popper's treatment of determinism per se; rather, the topic of determinism is addressed solely to the extent it evidences how Popper understood his philosophy to provide a solution to the metaphysical problem of induction. To rehearse what is at issue consider that the metaphysical problem of induction raises the question of how it is possible to talk about the structure of reality from within experience. To identify that a true natural law exists is to say something about the structure of the real. The question is whether it is possible to talk about the structure of reality from within experience in a manner consistent with scientific rationality. Hume argues that it isn't possible to justify in a rational manner even one natural law from experience, yet Hume thinks it within his rights to be a determinist concerning the physical world. Thus concerning the metaphysical problem of induction Popper writes:

> Hume's logical argument against induction simply does not immediately bear upon our metaphysical assertion that there exist regularities in nature. Nevertheless, it is perfectly true that we shall have to defend this metaphysical assertion against Hume – but not against his *logic*; rather, against his *metaphysics*.[130]

Consistent with Hume's logical critique of induction Popper concludes it isn't possible to establish a metaphysical thesis based on science since the former has a greater scope, but Popper's goal is to identify how scientific claims function as a critical check on metaphysical conjectures, yet at the same time a structural claim can be shown to make possible scientific testing.

Popper understands the metaphysical and scientific problems of determinism to be distinct because the former is devoid of empirical content and so is logically weaker than the latter, yet for this reason the metaphysical problem is entailed by it. 'Thus metaphysical determinism is, because of its weakness, entailed by ... "scientific" determinism; and it may be described as containing what is common to the various deterministic theories'.[131] This logical relationship between these two types of determinism serves to indicate the procedure Popper adopts to solve the problem of metaphysical determinism, and thereby speaks to as well the question of the structural properties of the universe, that is, the metaphysical problem of induction. He writes,

> It [metaphysical determinism] is irrefutable just because of its weakness. But this does not mean that arguments in its favour or against it are impossible. The strongest arguments in its favour are those which support 'scientific' determinism. If they collapse, little is left to support metaphysical determinism.[132]

Popper's argument against scientific determinism is as follows. Scientific determinism is 'the doctrine that the structure of the world is such that *any event can be rationally predicted, with any desired degree of precision, if we are given a sufficiently precise description of past events, together with all the laws of nature*'.[133] Scientific determinism received its greatest impetus from both Newtonian dynamics and Laplace's treatment of probability. According to Laplace, if there exists a demon who knows all the initial conditions and systems of natural laws for a limited region of space at a particular definite instant of time, then the demon could predict with exact precision the future course of any event in the specified spatio-temporal region. Now Popper argues that such a limited spatio-temporal region 'may be said to be "simultaneous" in the sense of special relativity'.[134] However, if we try to introduce Laplace's demon into special relativity, we find that the demon is only able to make retrodictions, and so the future is open. Popper argues this point as follows.

Consider a cut through a horizontal presentation of a Minkowski light cone such that the asymmetry between the past and the future proceeds from left to right. The apex of the light cone is designated by a point labeled *A*, and indicates the here and now. Let *B* represent an event in the future of *A*. Furthermore allow that a line segment *C* designates the limited spatio-temporal region of which the Laplacean demon has complete knowledge of all the initial conditions.

Diagram:[135]

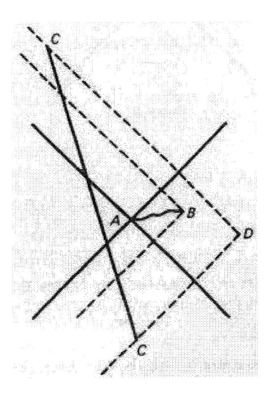

Now, according to Popper, the theory of special relativity,

> allows us to find a spatio-temporal position D which, from the point of view of the theory, is the *earliest* spatio-temporal position at which the demon may be located when receiving the information. And D will be so located that B belongs to the past of D.[136]

Thus Popper concludes the theory of special relativity refutes scientific determinism by requiring that Laplace's demon can perform only retrodictions. The future is decidedly open, and this is a feature of reality itself.

Keuth challenges the merit of the above argument for indeterminism asserting: 'this strategy of reasoning is a failure'.[137] Basically, Keuth's criticism is that Popper's argument proceeds from epistemic considerations to ontological/metaphysical conclusions. However, Keuth's critique is based on two fundamental errors. First, Keuth asserts a logical inference that Popper completely rejects. Keuth states:

> According to its form, the special theory of relativity is deterministic. If it should be *true*, the events to which it refers would be determined. This would *support metaphysical determinism*, which Popper wants to undermine. ... He does not, however, doubt the truth of special relativity, but he considers it "not *prima facie* deterministic," because, when we predict an event, we can never know all the other events that might possibly influence the event to be predicted. This, he thinks, *weakens "scientific" determinism* (the future is epistemically closed), and if it fails, *the most important argument supporting metaphysical determinism* (the future is ontologically closed) *is dropped*.[138]

But in no way does Popper accept the notion that an inference from the prima facie deterministic character of a scientific theory can support an inference to a theory of greater scope.[139] Indeed, such an inference is overtly inductive. Second, Keuth equates the notion of predictability with epistemic states. But although Popper talks about predictability and knowledge from within a physical system in the example of Laplace's demon, the point of the example is not per se to treat Laplace's demon as an observer in the sense of focusing on epistemic states; in fact, by 'observer' Popper prefers the expression 'local inertial system',[140] and what is characteristic of the system in the demon argument is that all initial conditions and systems of natural laws are present to allow for the deduction of the future state of the system having any desired degree of precision.[141] Finally, scientific determinism is itself a nightmare for physical theory because it entails that all probabilistic scientific theories are to be interpreted subjectively, that is, because the universe operates according to deterministic laws it follows that any probabilistic theory can only express a subjective measure – the ignorance of the scientist. However, Popper argues for an objective interpretation of probability and he is motivated to do so by the aim of removing the emphasis on the observer in the Copenhagen interpretation of quantum physics.

By the above refutation of scientific determinism, Popper demonstrates that a theory that refers to the structural properties of the physical universe can be classified as false and thus constrained by experience. Moreover, implicit in the above argument is the claim that indeterminism has important consequences for any account of the structural properties of the physical universe because 'indeterminism may be able to exclude certain possible arrangements of matter and motion in the world'.[142] Popper's argument that conjectures concerning structural properties of the physical universe are criticizable has opened the door to the classification of such statements as true or false, insofar as they withstand attempted refutations from within physical theory. Thus his solution to the metaphysical problem of induction entails that metaphysical claims and problems, such as Newton's problem, are neither simply an impetus to objective science as Popper himself once contended, nor meaningless as the logical positivists concluded. Instead, conjectures concerning the structural properties of the physical universe can be criticized, and either rejected as false or retained as true as part of the ongoing project of critical inquiry that characterizes physical science. But most importantly for scientific method, metaphysical theories, in this case indeterminism, help to make science possible:

> So far I have criticized determinism by trying to show its disadvantages ... [P]erhaps the strongest positive argument in favour of indeterminism: in rejecting determinism, we open the way for an approach that could be of real significance for science. I have in mind a physical interpretation of probability theory in the form of a physical theory of propensities. Even if such a theory, after serious discussion, ultimately proves unacceptable, the fact will remain that only by discarding determinism do we gain the freedom necessary for a serious consideration of the propensity interpretation as a physical theory. Thus determinism is not only unsupported by argument; it prevents us from seriously considering possibilities ...[143]

For those clerics and professors who refused to look through Galileo's telescope, metaphysics and 'empirical science' are mutually supportive in a way that precludes critical inquiry. But what Popper has shown in the name of scientific method is that empiricism and the metaphysical thesis of indeterminism are interdependent as part of the methodology of conjectures and refutations. In a rigidly deterministic universe criticism would proceed within very narrow limits and would be for the most part worthless, since in such a reality what does happen can't be otherwise; therefore, rationality and the growth of knowledge would be authoritarian and dogmatic, because ideas would have no choice but to be accepted. Thus Popper is a metaphysical realist about indeterminism and this structural claim about the universe is a boon to scientific rationality. That being said, again I wish to emphasize that the weighty debate concerning determinism is not taken to be settled; instead, what I hoped to have illustrated is how Popper thought that he could tackle the metaphysical problem of induction in a manner that merged metaphysical realism and empiricism. Finally, the above-mentioned theory of propensities understood as a generalized theory of forces is the central component to Popper's cosmology and intimately linked to his metaphysical worldview. But before turning to that topic as part of a survey of Popper's objectivist cosmology I want to address some essential corollaries to his account of scientific method.

3. Corollaries: Corroboration, Truth, and Verisimilitude

3.1. Corroboration

Popper's theory of corroboration addresses the topic of theory preference. His treatment of corroboration is motivated by 'the wish to grade hypotheses according to the tests passed by them', and he proposes 'to call the grade of a hypothesis, or the degree to which it has stood up to tests, its "*degree of corroboration*"'.[144] The intuitive idea is that inquirers classify as true those scientific statements that have been most severely tested. However, Popper adds two qualifiers that seem to mitigate the relevance of his theory of corroboration. First, he doubts that 'a difference of opinion regarding the acceptability of a hypothesis will ever be removed by an "exact" determination of its degree of corroboration'.[145] In part, exactness is problematic because corroboration is a complicated notion based upon other factors that allow for limited precision. Its logical development is as follows. The degree of corroboration is the result not simply of the number of tests to which a theory has been put, but it is an expression of the severity of the tests; however, the severity of tests 'depends upon the degree of testability, and thus upon the simplicity of the hypothesis: the hypothesis which is falsifiable in a higher degree, or the simpler hypothesis, is also the one which is corroborable in a higher degree'.[146] Now Popper asserts concerning empirical statements that results concerning their degree of testability are the same as the derivability relations (i.e., logical content) that holds between them.[147] But as Keuth correctly points out, 'we can seldom compare the testability of theories in this way, because deducibility relations are rare even among theories having the same subject matter'.[148]

Popper's second reservation concerns the practical significance of his definition of corroboration. He doubts that any numerical evaluation of his definition will ever help in evaluating the comparison of rival theories, for example, Einstein's and Newton's.[149] Given these two qualifiers why then the interest in corroboration? Popper recognizes that scientific testing is aimed at an appraisal of scientific theories, and he recognizes the obvious tendency to identify the extent to which a theory has passed testing with a probability measure where the latter is interpreted inductively. The idea is that the more tests a theory has withstood the greater its probability, thus testing could be understood as providing inductive support for a theory, hence the belief in inductive probability.[149] Popper is at pains to differentiate corroboration from the notion of logical probability because the former is an expression of empirical content, and given the inverse relationship between probability and content the higher a theory's probability the less it says about the world. Obviously, if testing led to an increase in the support of a theory in the sense of inductive probability, then testing would undermine the explanatory power of science because it would provide less information about the world. The debate over the success of Popper's attempt to dissociate corroboration from probability is ongoing. However, what needs to be better understood is that inductive probability is an epistemic notion while corroboration 'has no epistemological significance at all, as Popper always insisted'.[150] The point of inductive probability is to establish the likelihood of a theory as true, to justify it, and thus to satisfy epistemological aims. In contrast, the corroboration of a theory is a report about testing, but a report is not an inference, so corroboration is in no way ampliative in the way inductivists understand inductive probability to be. Indeed, that something is reported to be true is no supporting reason to believe it is true or will be true tomorrow. Only the reports of sources given an antecedent status as an authority (e.g., for some people the evening news) have epistemological significance. Scientific tests are at the heart of scientific activity and are undertaken to help scientists discover the truth about the world, but a report about the severity of tests a scientific theory has undergone does not support the inference that the theory will not show itself to be false in the future. Additionally, if the aim of science is to attain high probabilities for scientific theories based upon testing, then, Popper argues, the result is timidity not characteristic of scientific advance:

This philosophy [inductive probability], in regarding it as the aim of science to attain high probabilities for its theories, implies that science proceeds according to the rule: "Go as little as possible beyond your evidence e!" For the content of our hypothesis h cannot go very far beyond the evidence e without reducing $p(h, e)$ to a value very close to zero. For example, let e be the conjunction of many descriptions of an event of a certain kind: h does not need to assert many further events, not covered by e, in order to make $p(h, e)$ very small, according to the calculus of probability. This shows that a high probability is the dubious reward for saying very little, or nothing. In other words, the rule "Obtain high probabilities!" puts a premium on ad hoc hypotheses.[151]

3.2. Truth

It is well known in Popperian circles that in his landmark *The Logic of Scientific Discovery*, Popper is reluctant to address the topic of truth. On page 273 of *Logic* he writes, '[i]n the logic of science here outlined it is possible to avoid using the concepts "true" and "false"'. As he later reveals, his early reluctance to address the topic of truth in a more confident manner is grounded in the 'difficulty of explaining the correspondence theory: what could the correspondence of a statement to the facts be?'[152] Popper asserts that Tarski's theory of truth helped him to overcome his tergiversations concerning truth, but as Kirkham points out, it is still a matter of intense debate whether Tarski's theory should be understood as a correspondence theory.[153] Leaving this very detailed issue to the side I want to address what I understand to be the elements in Tarski's theory of truth that Popper found most valuable and how these elements contribute to his treatment of objectivity.

Tarski's theory of truth aims to provide a materially adequate and formally correct definition of 'the notion of truth'.[154] Scientific semantics asserts both an extension and an intension to the terms it investigates. The extension of 'truth' is sentences, although this doesn't exclude an extension to other types of objects.[155] For semantics the problematic aspect of the term true is its intension, that is, what are the list of attributes that comprise an adequate definition. The 'familiar formula: the truth of a sentence consists in its agreement with (or correspondence to) reality' (ibid.) is unclear, imprecise, and leads to antinomies such as the liar's paradox. The material adequacy and formal correctness conditions in conjunction with a distinction between levels of language (object level and meta-level), paralleling Russell's theory of types, are meant to rectify this situation.

Minimally, every true sentence asserts an equivalence of the kind represented in the following example: '"Snow is white" is true if and only if snow is white.'[156] Here the expression in double quotation marks signifies the name of the sentence on the right-hand side of the equivalence. Generalizing from the example, Tarski identifies an ' "*equivalence of the form T*" ' or T-scheme: X is true if, and only if, p wherein: 'p' is a sentence and 'X' is the name for it. A definition of truth is materially adequate for a given language if all instances of the T-scheme deductively follow from the definition. Tarski goes on to add that neither the T-scheme nor 'any particular instance of the form (T) can be regarded as a definition of truth'.[157] This follows because the T-scheme, not being a sentence, and thus neither true nor false, can't entail the truth of anything; as well, by replacing 'p' and 'X' with appropriate respective instances the result is only a 'partial definition of truth'.[158] A complete definition requires the 'logical conjunction of infinitely many sentences'.[159] Semantics, in one sense, concerns the relation between expressions and the objects they refer to. Tarski identifies designation, definition, and satisfaction as relations of fundamental concern to semantics; indeed, the latter is central to his definition of truth. Truth, however, is logically distinct from these relations, 'it expresses a property (or denotes a class) of certain expressions, viz., of sentences'.[160] Because a language sufficiently rich for, say, mathematics denotes a class containing infinite members its extension can't be realized; consequently, a technique for an intensional account of truth is required. If both the material adequacy and the formal correctness conditions are to be fulfilled in the name of an adequate definition of truth, then, as Miller points out, the material adequacy condition must be restricted so that instances of the T-scheme are derivable for the object language.[161] Thus the distinction between an object language and a meta-language is a consequence of avoiding semantically closed languages since such languages allow for inconsistencies, and hence the antinomy of the liar.[162] A semantically closed language contains not only expressions, their names, and 'the term "*true*" referring to sentences of the language; we have also assumed that all sentences which determine the adequate usage of this term can be asserted in the language'.[163] Truth *applies to* sentences of the object language while '[t]he definition itself and all the equivalences implied by it are to be formulated in the metalanguage'.[164] For example, if the French sentence 'Il pleut' is the object language and the English sentence 'It is raining' is the meta-language, then to avoid semantic closure such that the object language is not plagued by antinomies, truth is defined at the meta-level. An adequate meta-language sufficient for a definition of truth is logically richer than the object language but since, like the object language, it has within its domain open sentences that are neither true nor false, truth can't be defined by building up in a piecemeal fashion a comprehensive and sufficient definition based upon the truth of the component elements of the meta-language per se. Thus truth is defined in reference to another property, satisfaction, which applies to both open and closed sentences. The open sentence 'x is white' is satisfied by snow and thus true, and the closed sentence (i.e., an open sentence that has been quantified) '(x) (x is white)' is true *'if it is satisfied by all objects and false otherwise'*.[165]

For Popper the first and most important aspect of Tarski's definition of truth is that it can be interpreted as commensurate with realism. Here the operative word is 'can', because contrary to what Keuth asserts[166] Popper recognizes that Tarski understands his theory to be neutral concerning various epistemological stances.[167] For Popper, Tarski's theory allows for a realist interpretation commensurate with the growth of scientific knowledge in the following way. Popper understands Tarski's T-scheme to render truth eliminable because 'p' in the T-scheme describes a state of affairs that corresponds with reality. Thus rather than say ' "Snow is white" is true', one can replace the expression 'is true' with the sentence 'Snow is white.' Nevertheless, this eliminability is grounded in more than a mere synonymy that simply allows the truth predicate to cancel 'linguistic reference';[168] rather, I believe that Popper applauds the eliminability because the T-scheme evidences an extensional equivalence that is relevant to scientific practice. The identification of extensionally equivalent descriptions is often a characteristic of great scientific insight; for example, Maxwell's account of light as electromagnetic radiation. Popper is adamant that genuine scientific and philosophical work is not concerned with the meaning of words and so the issue of truth is about more than just a certain utility associated with linguistic reference. This is why he reproves any attempt to eliminate philosophical or scientific problems by an a priori determination of when a reduction/synonymy is acceptable and when it is not.[169] Thus Popper sees Tarski's account of truth in terms of the satisfaction of a sequence of objects to be amenable to realism.

As well, on Popper's reckoning Tarski's theory of truth enables truth to fulfill its regulative function.[170] The liar's paradox precludes truth from fulfilling its regulative role since it presents us with a statement that is true if it is false and false if it is true. Without this regulative function, especially in the realist sense of truth as correspondence with the facts, science as the method of conjectures and refutations would be impracticable because experience would be denied its selective function. Finally, Tarski's theory shows that there can be no criterion of truth, since what truth is and how we come to establish something as true are distinct. This, as Miller rightly points out, has the liberating effect of removing the need to define truth 'in terms of the procedures we use to classify it'.[171] Because epistemology since the

Renaissance has concerned itself with the pursuit of a criterion of truth in the name of foundationalism along rationalist or empiricist lines, Tarski's theory of truth aids Popper's anti-foundationalism. Moreover, the equivalence characteristic of the T-scheme makes no reference to belief states and so truth understood as having a privileged relationship to the knowing subject is eliminated.[172]

3.3. Verisimilitude

The idea of objective progress to the truth is important to scientific activity. Scientists, in a Popperian world, criticize their conjectures in the hope that they can eliminate what is false in their theories and so get closer to the truth. However, giving a precise account of this very intuitive idea of getting nearer to the truth or verisimilitude has shown itself to be very difficult. Popper invokes the idea of verisimilitude to explain the progress of science.[173] His account of scientific progress incorporates two fundamental ideas: (1) truth, and (2) content. 'Truth' is understood to denote correspondence. 'Content' refers to testability and consists of two types: logical and empirical. The phrase 'logical content' denotes the class of all statements that follow logically from a given statement A. 'Empirical content' refers to the class of all statements which contradict a particular theory. Working from the above taxonomy Popper defines verisimilitude as the truth content of a theory A minus the falsity content of the theory. Schematically:

$$Vs(A) = Ct_T(A) - Ct_F(A)^{174}$$

However, as was demonstrated independently by Miller and Tichy, this definition is viciously language dependent. The refutation, readily acknowledged as such by Popper, proceeds in the following way.[175] Consider a theory consisting of the conjunction of the statements A & B & C to be true. Next consider two false theories $T1$ and $T2$: where $T1$ asserts $\sim A$ & B & C, and $T2$ asserts $\sim A$ & $\sim B$ & $\sim C$. Obviously, given the assumed truth of the conjunction A & B & C, $T1$ is closer to the truth than $T2$. However, this verisimilitude is not maintained upon translation. A new theory can be constructed which includes A, but also two other statements defined as follows: (1) D df $= A$ iff B; (2) E df $= A$ iff C. Given the truth of the new theory comprising the conjunction A & D & E, $T1$ now reads $\sim A$ & $\sim D$ & $\sim E$ and $T2$ is translated as $-A$ & D & E. Thus the verisimilitudes of the false theories have switched. Given that this argument demonstrates the language dependence of Popper's account, it is difficult to understand how judgments of verisimilitude can correspond to some objective aspect of the world.[176]

The problem is not trivial and will not go away by asserting that all theories incorporate undefined terms, and verisimilitude is to be adopted as such.[177] Although Popper claims correctly that his definition of verisimilitude is a metalogical concept and not fundamental to his methodology, the notion of a better approximation to the truth is important to an understanding of both objective science and objective scientific progress.[178] First, scientists employ judgments of verisimilitude all the time, for example, when they conclude that Darwin's theory of evolution is closer to the truth than Lamarck's. Second, judgments of verisimilitude effect the direction of scientific inquiry because for the most part scientists investigate those theories that they believe to be closer to the truth. Thirdly, a coherent explanation of verisimilitude involves the idea of correspondence which underlies, in part, Popper's metaphysical realism and hence his treatment of objectivity. Lastly, what does the lack of a coherent account of verisimilitude imply for the notion of truth as a regulative ideal: what is it that truth is regulative of, if not truthlikeness?

I conclude with Miller that an intelligible account of objective judgments of verisimilitude is not yet at hand.[179] Some Popperians have attempted to rescue the notion of verisimilitude by divorcing it from the ideas of truth content and falsity content. On their reckoning, the general notion of verisimilitude may be rendered an intelligible, that is, criticizable feature of science, if it is transfigured via evolutionary epistemology as the idea of a better fit between theory (adaptation) and fact (environment).[180] Although such attempts are interesting in themselves, they are not fit to the challenge.

Popper's treatment of verisimilitude was intended to explain how objective judgments of degrees of truthlikeness between theories could be made. Verisimilitude was conceived as a metalogical concept intended to explain scientific progress. The notion of a better fit between theory and fact, however, is akin to the degree of corroboration of a theory and corroboration is the methodological counterpart of verisimilitude.[181] Thus any identification of verisimilitude with the notion of a better fit between theory and fact conflates the metalogical with the methodological. And because Popper states that the failure of his definition of verisimilitude has no influence on his account of scientific method, it follows that the two are not to be identified.[182] A better testable, i.e., corroborated theory, will fit with the current state of the available scientific facts, but that a theory is better testable does not entail that it is more true. Moreover, underlying the difference between verisimilitude and the idea of a better fit between theory and fact is the distinction between the growth of a scientific theory and scientific progress.[183]

Science progresses when it conjectures new theories that are indeed true. Science can be said to grow, however, when a theory is shown to fit, for example, with the established conclusions of subordinate disciplines; however, the fact of this fit does not manifestly add to the progress of scientific truth. The growth of a scientific theory and scientific progress are distinct.

For Popper, science is characterized by the elimination of false hypotheses from the domain of science. And to the extent that we are successful at the elimination of false hypotheses our knowledge grows. However, we should not conclude that because we can eliminate false hypotheses, and because we now know it is true that a falsified hypothesis is not a part of science, that we have added to the truth content of science. Science does not attain more truths by simply identifying all the statements that are unscientific. Any positive aspect of the notion of verisimilitude is trivialized if the identification of such statements is all that it is intended to establish.

The present state of the philosophical discussion concerning verisimilitude forces us to conclude that it is an incoherent but desired appendage to Popper's account of scientific method. In Popper's philosophy the concept of verisimilitude functions in an as yet unclear way as an unclear regulative ideal. Judgments of verisimilitude may correspond to an objective state of affairs but to date a clear account has yet to be provided of how to measure an intuitively appealing idea. However, it is nevertheless important to realize that Popper's failure to provide an adequate account of this concept affects only the notion of objective scientific progress and not his account of objectivity per se. Popper's technical account of verisimilitude was intended to explain objective scientific progress. At first glance, that he cannot do so appears greatly disparaging to his account of objectivity. However, it is criticizability as a methodological constraint directed at the content and logical form of our conjectures that demarcates the objective from the subjective. Verisimilitude is not a methodological concept although it is an accouterment to scientific methodology whose clarification is worth pursuing.

The tendency to think that the failure of his treatment of verisimilitude egregiously undercuts his account of objectivity arises because philosophers have always believed that the idea of truth is central to any account of objectivity and thus the same holds for verisimilitude as well. But for Popper this is not entirely the case. I point out in the next chapter that Popper's World Three as the locus of objectivity contains false statements whose objectivity is secured by their criticizability and their ability to affect causally the world around us. Nevertheless, along broad lines an adequate account of verisimilitude would help in the battle against what Popper calls historism or today we might characterize as the strong program in the sociology of knowledge. For the proponents of the strong program in the sociology of knowledge and their post-modern cousins, progress is taken to be an Enlightenment chimera predicated on the humbug of objective standards of progress. Thus a coherent account of verisimilitude might do much to dissuade those thinkers who are trying to dissuade culture in general concerning the objectivity of science.

Notes

1. Popper (1959a) 15.
2. Popper *et al.* (1974) 976–7.
3. Popper (1959a) 422–3.
4. Ayer (1952) 153.
5. Popper (1959a) 313.
6. Ibid. 34.
7. Popper (1983) 161.
8. Ibid. 159.
9. See Bartley (1984) 113.
10. Popper (1959a) 30.
11. Ibid. 39.
12. Ibid. 41.
13. Popper (1983) xxii.
14. Popper (1959a) 102.
15. Ibid. 82.
16. Popper (1963) 218.
17. Popper (1983) 21.
18. Popper (1959a) 80.
19. Ibid. 79.
20. Ibid. 80.
21. Ibid. 39.
22. Popper (1983) 13, and Miller (1994) 28, 119.
23. Popper (1959a) 47.
24. Miller (2006) 108–9 disagrees with Popper's treatment of the Quine/Duhem problem.
25. Popper (1983) xxii.
26. Popper (1959a) 44
27. Ibid. 45.
28. Popper (1990) 27–51.
29. Bloor (1976).
30. Although what is at issue has invoked a weighty philosophical literature, the main point briefly can be put as follows. The conjunction of the two conditionals 'If something is possible, then it is conceivable' and 'If something is conceivable, then it is possible' is the biconditional $P \equiv C$. $P \equiv C$ is logically equivalent to $\sim C \equiv \sim P$; however, the latter biconditional is patently false whether conceivability denotes a property of a knowing agent or the content of a proposition. Because a person

can't conceive God's omniscience or Heisenberg's Indeterminacy Principle it doesn't follow that either is impossible. Moreover, to identify conceivability with the unique content of a proposition requires some constraint on what content is to be accepted as conceivable. Traditionally, the law of non-contradiction is understood to satisfy this demand. Most often, this constraint is approached from the side of possibility, since a contradiction identifies what is impossible; however, given the presumed relationship between conceivability and possibility if a state of affairs is contradictory it should be inconceivable as well. But what is it about the law of non-contradiction that enables it to suffice as a constraint on conceivability? At first glance, it may be thought that because contradictions are necessarily false statements one can't conceive them. However, in a reductio proof a person must understand the contradiction if she is to understand the proof, that is, a person has to grasp the necessary falsity that follows from making the negation of the conclusion a premise in the argument if she is to grasp the truth of the conclusion as a deductive consequence of those premises. Thus it isn't the property of necessary falsity that precludes conceivability. For more on this issue see Tidman (1994) and Gendler and Hawthorne (2002).

31. Popper (1959a) 449.
32. Ibid. 38.
33. Popper (1957).
34. Miller (1994) 106.
35. Miller (2006) 84.
36. Ibid. 85.
37. Ibid.
38. Miller (1994) 120.
39. Popper (1983) xxv.
40. Jarvie (2001) 18.
41. Ibid. 12.
42. Ibid. 14.
43. Ibid. 18.
44. Popper (1963) 207.
45. Jarvie (2001) 20.
46. Ibid. 21.
47. Curie (1938) 167.
48. Jarvie (2001) 47.
49. Popper (1959a) 29.
50. Ibid. 41.
51. Ibid.
52. Binns (1978).
53. Popper (1959a) 90, and (1983) 74, 182.
54. Popper (1959a) 84.
55. Ibid. 94–5.
56. Ibid. 93.
57. Ibid. 94.

58. Ibid.
59. Ibid. 95.
60. Ibid. 97.
61. Ibid. 105.
62. Ibid.
63. Popper (1983) and (1990).
64. Popper (1959a) 102.
65. Ibid. 103.
66. Ibid.
67. Popper (1979) 42–3. Here Popper presents a very interesting argument against idealism provided by Winston Churchill.
68. Popper (1959a) 104. Quoted in Keuth (2005) 100–101. Italics are Keuth's.
69. Keuth (2005) 101.
70. Ibid. 102.
71. Popper (1959a) 104.
72. Miller (1994) 48.
73. Hesse (1974) 95, quoted in Miller (1994) 18.
74. Popper (1983) 297.
75. Miller (1994) 31.
76. Keuth (2005) 114.
77. Deluty (2005) 87.
78. Kripke (1982) 5.
79. Ibid. vii.
80. Ibid. 146, footnote 87.
81. Ibid. 60.
82. Wittgenstein (1968) 81, para. 201.
83. Kripke (1982) 69.
84. Ibid. 11–12.
85. See Fitch (2004) 153–8 for an excellent exposition of the central issues.
86. Kripke (1982) 112.
87. Ibid. 99.
88. Ibid. 146, footnote 87.
89. Popper (1963) 390.
90. Boon (1979), quoted in Miller (1994) 33.
91. Miller (1994) 28.
92. Popper (1959a) 30.
93. Musgrave (1999) 333.
94. Keuth (2005) 125. Italics are Keuth's.
95. Popper (1983) 54.
96. Popper (1959a) 30.
97. Miller (1994) 5.
98. Popper (1983) 54.
99. Ibid. 56.
100. Ibid. 19.

101. Ibid.
102. Popper (1957) 13.
103. Musgrave (1999) 333.
104. Ibid. 335.
105. Ibid.
106. Ibid. 332.
107. 'Evolutionary Computation' Natural Selection Inc. selection.com/tech_1.html. Accessed 8/7/2006.
108. Popper (1959a) 17. Also, see Miller's own rejoinder, Miller (2006) 128ff. (For a related point see section II.1 above.)
109. Popper (1983) 63.
110. Ibid. 66–7.
111. Ibid. 65.
112. Popper (1959a) 30.
113. Ibid.
114. Popper (1983) 11.
115. Keuth is the exception. See Keuth (2005) 124–32.
116. Popper (1983) 80.
117. Ibid. 149.
118. Ibid. 161.
119. Ibid. 76.
120. Ibid. 82.
121. Ibid. 157, footnote 1.
122. Popper (1963) 193–200.
123. Ibid. 195.
124. Ibid.
125. Ibid 198. Italics are Popper's.
126. Ibid. 199.
127. Ibid.
128. Miller (1994) 11.
129. Popper (1983) 88.
130. Ibid. 76. Italics are Popper's.
131. Popper (1982a) 8.
132. Ibid.
133. Ibid. 1–2.
134. Ibid. 60.
135. Ibid.
136. Ibid. 60–1.
137. Keuth (2005) 279.
138. Ibid. Italics are Keuth's.
139. Popper (1982a) 38.
140. Ibid. 57.
141. Ibid. 31.
142. Popper (1982b) 205.

143. Popper (1982a) 93.
144. Popper (1983) 220.
145. Ibid.
146. Popper (1959a) 267.
147. Ibid. 122.
148. Keuth (2005) 117.
149. Popper (1983) 221.
150. Miller (1994) 120.
151. Popper (1983) 222–3.
152. Popper (1979) 320.
153. Kirkham (1997) 170.
154. Tarski (1944) 48.
155. Ibid. 49.
156. Ibid. 50.
157. Ibid.
158. Ibid.
159. Ibid.
160. Ibid. 51.
161. Miller (2006) 172.
162. Tarski (1944) 54.
163. Ibid. 53.
164. Ibid.
165. Ibid. 56.
166. Keuth (2005) 142.
167. Popper (1979) 323.
168. Quine (1970), quoted in Keuth (2005) 144.
169. Popper (1979) 292–4 and (1982a) 162–74.
170. Popper (1979) 317–18.
171. Miller (2006) 176.
172. Popper (1963) 225.
173. Ibid. 228ff.
174. Ibid. 234.
175. This presentation of Miller and Tichy's argument is taken from Eric Barnes (1995). I understand Barnes' analysis to be a good general assessment of what is common to Miller and Tichy's papers.
176. For Miller's current assessment of the debate concerning verisimilitude, see Miller (2006) Chapter 11. Here, Miller downplays structuring the problem of verisimilitude as primarily a problem of language dependence. However, he argues 'the threat of subjectivism and relativism still lurks nonetheless.... The outstanding problem then seems to be the problem of how different respects [ways of approaching the truth of a state of affairs] are to be objectively weighed or aggregated, how overall similarity is to be assessed' (232).
177. Popper (1983) xxxv–xxxvii.
178. Popper (1963) 235.

179. Miller (2006) 233.
180. Bartley (1987).
181. Popper (1963) 235.
182. Ibid.
183. Miller (1994) 199.

Chapter 3

Cosmology and Propensities

Introduction

'[W]e live in *a world of propensities*'.[1] What exactly does this mean and how is such a claim reconcilable with objectivity? Popper's claim is meant comprehensively and this entails that the theory of propensities should extend systematically to his entire philosophy. Popper never works out the implications of this idea for all the various particular aspects of his thought, but neither does he run away from the metaphysical implications of this claim. Indeed, his theory of propensities can be studied from the perspective of its extension to his World Three ontology that posits: World One, the world of physical objects and their interactions; World Two, the world of human conscious activity; and World Three, the products of human conscious activity such as theories, arguments, and mistakes, understood as a realm of statements-in-themselves. In what follows, I adopt this division as a resource to explore Popper's theory of propensities without necessarily arguing for such an ontology as a praiseworthy element of Popper's philosophical outlook. Consistent with the above ontology, in what follows I present an overview of Popper's theory of propensities as a physical theory, and next argue for evolution in general and human evolution in particular as a propensity that gave rise to what Popper understands as an objective domain of statements-in-themselves that is itself characterized by an openness to or propensity for criticism that causally affects the world.

1. Popper's Account of Propensities

Popper's *A World of Propensities* contains his most mature formulation of his propensity theory as well as an indication of its extension into other areas of his philosophy. Popper's propensity theory is often discussed only in conjunction with his work on probability. And in fact, Popper rarely formulates his general theory of propensity without reference to his objective formulation of the probability calculus. This fact has contributed to the propensity theory not being understood as a physical theory. But his theory of physical propensities, while apiece with his objective interpretation of the theory of probability, is not simply an instrument to aid mathematical calculation.

Physical propensities are analogous to forces and fields of forces. Popper states:

> Propensities, it is assumed [i.e., conjectured] are not mere possibilities but are physical realities. They are as real as forces, or fields of forces. And vice versa: forces are propensities. They are propensities for setting bodies in motion. ... Fields of forces are fields of propensities. They are real, they exist.[2]

Popper's propensity theory emphasizes the physical situation. However, propensities are not properties inherent in physical objects as such, but instead they are properties inherent in an entire physical situation, which may either change or remain constant.[3] The propensity of a person to live another year is not just a question of her state of health understood as denoting a property of her. Instead, the propensity that she will live another year is dependent upon other factors that comprise the entire physical situation; for example, whether new disease-fighting drugs are developed, or whether there is a dramatic increase in violent crime.

Physical propensities are an extension and generalization of the idea of forces. As such they can be understood as generalized causes. Accordingly, a cause having produced an effect – 'the case of a classical force in action' – is an instance of a definite propensity to which we ascribe the numerical value 1.[4] A propensity having a numerical value less than 1 but not zero is to be understood as analogous to opposed competing forces which at present do not control a physical process. No propensity, to which we ascribe the numerical value 0, is present when both no cause and no effect exist to characterize a physical situation. For example, there is no possibility, and thus no propensity, to obtain the number 14 with the throw of two fair die.[5] Thus, for Popper causal laws and causal explanation, which have been traditionally understood to be the special province of empirical science, are now best treated as unique limiting cases of probabilistic laws and probabilistic explanation.[6] Propensities are themselves unobservable but they have testable physical properties that are measurable.[7] According to Popper, an excellent example of a physical propensity that can be measured is the half-life of a radioactive nucleus.[8] Indeed, on Popper's reckoning, Rutherford and Soddy's account of radioactive decay is only understandable in reference to the conjecture 'that each atomic nucleus had a *tendency or propensity* to disintegrate, dependent upon its structure'.[9]

In sum, physical propensities are the effect of a structured, relationally defined physical situation in which particular possibilities are realized in a way that renders them testable. Now since *every* physical event realizes a conjoining of various physical possibilities culminating in structural properties that give rise to a propensity, it follows that every physical event is both a propensity to realize a particular structural arrangement of physical possibilities and the effect of a propensity having a particular structure. As Popper puts it:

> According to this picture ... all properties of the physical world are dispositional, and the real state of the physical system, at any moment, may be conceived as the sum total of its dispositions – or its potentialities, or possibilities, or propensities. Change, according to this picture, consists in the realization or actualization of some of these potentialities. These realizations in their turn consist again of dispositions – or potentialities, though, that differ from those whose realizations they are.[10]

Consequently, if we treat the world as a hierarchy of propensities it seems to be the case that some propensities are both more fundamental than others and situationally more present. For example, the element hydrogen is much more present in the universe than the element cesium. And hydrogen seems to be more essential to a great many physical events; for example, in the formation of water essential for life and in the nuclear reactions within the interior of stars. And the fact that hydrogen atoms have a relatively stable structure is dependent upon their entire physical situation delimited in part, but not completely, by: (1) their configuration of protons and electrons, (2) the interplay of strong and weak nuclear forces which hold them together, and (3) the fact that the structure of the universe per se is not characterized by the thermo-nuclear chain reactions of stars in which hydrogen atoms are destroyed.

At the heart of Popper's theory of physical propensities are claims concerning the omnipresence of structural properties of the physical universe. These claims are parallel to Newton's assertions about the omnipresence of gravitational forces. Indeed, to those critics who charge that his physical theory of propensities introduces 'occult' or 'metaphysical' qualities into science, Popper is quick to point out in his own defense that the same charges were leveled against Newton's theory of forces. Like Newton's theory of universal gravitation, Popper's propensity theory makes assertions about the structural properties of the physical universe. However, Popper's theory is related to Newton's theory and all true physical theories of causation as a meta-theory. 'The introduction of propensities amounts to generalizing and extending the idea of forces again.'[11] Moreover the success of the analogy between the two ideas of force and propensity is to be found 'in the fact that both ideas draw attention to unobservable dispositional properties of the physical world, and help in the interpretation of physical theory'.[12]

1.1. Physical Propensities, Probability, and Tests

Popper's defense of his theory of physical propensities depends upon both its success in the interpretation of physical theory and the support it receives through an objective interpretation of the probability calculus. Thus far we have seen that Popper's defense of his theory of propensities has, in part, rested upon his assertion that as an empirical claim it is analogous to Newton's theory of forces. However his contention has been challenged. One such objection asserts that for Newtonian forces there are explicit formal laws which Newtonian forces must obey, but which Popper's propensity theory does not obey. For example, there are the laws of addition of forces.[13] In response to this challenge to the objective status of propensities Popper states,

> [T]here is an addition law for propensities, as there is one for forces. That this law is part of the probability calculus is perfectly true, but this is so because the calculus of relative probability (as axiomatized in my L.Sc.D.) is a good axiomatization of the general theory of propensity.[14]

Therefore to understand Popper's defense of his propensity theory, I present his objective interpretation of the probability calculus. To this end, I begin with an account of Popper's early theory of objective probability and the reasons for his rejection of this theory in favor of the propensity theory. Also I indicate how the propensity theory of probability can render probability statements falsifiable, and thus enable them to be treated as empirical scientific statements.

a. Popper's Early Treatment of Probability

Popper's treatment of probability was initiated by two fundamental problems. The first problem – directly related to his methodology – concerns his criterion of demarcation as presented in *The Logic of Scientific Discovery*. In that work Popper raises the question of whether probability statements are strictly falsifiable, and argues that they are not.[15] This follows because no predictions are strictly deducible from probability statements. Thus there is no way to deduce a false consequence from a probability theory; or, for that matter, to verify one. For example, the probability of obtaining a two with the toss of a fair die is unaffected by any sequence of throws because, in the long run, any deviation from the outcome of 1/6 can either occur or be corrected for. Moreover, Popper states:

> Probability hypotheses *do not rule out anything observable*; probability estimates cannot contradict, or be contradicted by, a basic statement; nor can they be contradicted by any finite number of basic statements; and accordingly not by any finite number of observations either.[16]

However, probability statements comprise a large number of the statements of science. And if probability statements are not falsifiable, then many of the statements that Popper and many scientists would want to accept as empirical scientific statements would have to be classified as non-scientific. To explain how probability statements are a part of empirical science, Popper makes it his aim 'to elucidate the relations between probability and experience. This means solving what I [Popper] call the *problem of decidability of probability statements*.'[17] The next problem Popper's early account of probability addresses is labeled the fundamental problem of the theory of chance. This problem raises the question of 'how a statement of ignorance, interpreted as a frequency statement, can be empirically tested and corroborated'.[18]

In *The Logic of Scientific Discovery*, Popper identifies three interpretations of probability. These are: (1) classical, (2) subjective (further subdivided into the psychologistic and logical), and (3) frequency. In solving the two fundamental problems of probability, his approach in *Logic* is further to develop the frequency theory.[19] Popper believes that the frequency theory of probability, understood as an objective interpretation, is the only interpretation of probability that can do justice to the empirical character of probability statements and the use made of them in physics.[20] As such, it stands in direct contrast to any subjective interpretation of probability statements. Interestingly, Popper has little to say about the classical or Laplacean account of probability even though this account of probability is later instrumental in the development of his propensity theory.[21]

However, in *The Logic of Scientific Discovery*, Popper identifies two instances of a subjective theory of probability. The first type of subjective interpretation is labeled psychologistic.[22] This interpretation treats probability statements as a measure of belief or doubt, i.e., a measurement of a person's level of ignorance regarding a particular state of affairs. The second instance of a subjective interpretation is the logical interpretation of probability statements. According to this interpretation, probability measurements express a 'special kind of logical relationship between two statements'.[23] In this interpretation probability statements indicate what Keynes called the ' "degree of rational belief" ', and they express logical relations like derivability, incompatibility, etc.[24] Popper readily admits that each subjective interpretation can delineate the probability calculus in a manner consistent with its subjective starting point. However, Popper denies that such interpretations 'describe the use of probability in the physical sciences'.[25] On his reckoning, such interpretations reduce probability statements to tautologies, and fail to tell us anything about the objective world which is the object of inquiry for the physical sciences.

In contrast, Popper argues that his version of the frequency theory is the best objective interpretation of probability statements. In effect his treatment is a reworking of the frequency theory of von Mises. Therefore, let us briefly consider von Mises' frequency theory to better understand Popper's emendation of it.

b. Von Mises' Frequency Interpretation

The frequency theory of probability initiated by von Mises in *Warscheinlichkeit, Statistik und Wahreit* characterizes probability statements as mathematical measurements performed on a collective.[26] The term 'collective' denotes random sequences of events capable in principle of being continued indefinitely.[27] A collective must satisfy two axioms: (1) the axiom of convergence, and (2) the axiom of randomness. Each of these axioms serves to qualify the notion of a relative frequency. A relative frequency is the number of favorable outcomes in a series of outcomes compared to the total number of outcomes, and can be expressed as a number from 0 to 1. For example, the relative frequency of obtaining a 2, given a sufficient number of throws with an honest die, is 1/6. In von Mises' frequency theory the axiom of convergence requires that an event sequence, expressed as a relative frequency, tends toward a definite limit. The axiom of randomness postulates that no gambling system, or Dutchbook, can be applied to a sequence or set of sequences for which a relative frequency is established. If the two axioms are satisfied, then, according to von Mises, a probability is the limiting relative frequency of an outcome relative to the collective.[28] Or as Popper puts it:

> The *task of the calculus of probability* consists, according to von Mises, simply and solely in this: to infer certain 'derived collectives' with 'derived distributions' from certain given 'initial collectives' with certain given 'initial distributions'; in short to calculate probabilities which are not given from probabilities which are given.[29]

c. Popper's Frequency Theory of Probability

Popper rejects the frequency theory of von Mises. He does so because his analysis of the role of both the axioms of convergence and randomness led him to conclude that the adoption of both axioms results in a paradox. The paradox arises because the mathematically derived idea of randomness is conjoined to the idea of a mathematically derived limit. Accordingly, he argues that the frequency theory can be salvaged by: (1) improving the axiom of randomness, and (2) completely eliminating the axiom of convergence. According to Popper the adoption of each of his emendations enables him to solve the two fundamental problems of probability.

Improving the axiom of randomness is fundamentally a mathematical problem. Popper's criticism of von Mises' account of randomness is focused on the latter's account of infinite sequences. According to Popper, von Mises argues that a sequence can be shown to be sufficiently random, that is, no gambling system can be constructed for it, only if we grant that, in the long run, any apparent gambling system that is applicable to a subsequence of a collective will eventually disappear. However, according to von Mises, it is possible that some segment of a collective may show great regularity for what, at least initially, appears to be a large segment of the collective. Popper labels this deference to the outcome of the sequence in the long run (i.e., deference to its potential infinity), 'deferred randomness'. Indeed, according to Popper, von Mises' frequency theory only excludes 'extremely regular' commencing segments of a collective, given the notion of deferred randomness.[30] And Popper asserts 'that we cannot empirically test *this* kind of deferred randomness'.[31] Popper's response to the frequency theory's untenable idea of deferred randomness is to demonstrate how to construct a random sequence whose randomness is evident for both long and very short commencing sequences.[32] The success of Popper's account of randomness enables him to eliminate the axiom of convergence, because the first law of great numbers is a logical consequence of any consistent notion of randomness.[33] And the first law of great numbers asserts that, given a random sequence, a limit appears whose probability approaches certainty (represented by the number 1) if the sequence is sufficiently long. Thus the axiom of convergence is no longer necessary. Moreover, Popper rejects the axiom of convergence because it is implicitly inductive. It postulates that a limit can be derived for a collective based upon samples taken from measures on a collective.

Given the elimination of the axiom of convergence, the fundamental problem of the theory of chance is solved without contradiction. Basically, this problem raises the question of how order is to be derived from a truly random process. As was demonstrated, once randomness for a sequence of any length is established, a limit for the sequence can be attained in accordance with the first law of great numbers. Furthermore, the problem of decidability is also solved by Popper's modified notion of randomness. Allowing that we can now ascertain if any particular segment of a sequence (i.e., a finite random sequence) is indeed random, it follows that we can determine the randomness of the sequence per se. If a random sequence cannot be constructed for a collective, then it cannot show itself to be random. Thus it cannot be an object of probability measurements. Consequently, its status as an event sequence able to generate probability statements is falsified.[34]

d. The Propensity Theory of Probability

Over time Popper became dissatisfied with all versions of the frequency theory of probability because of their inability to confront the problem of single case probabilities. The problem of the single case raises the question of the validity of assigning the relative frequency of the collective to a single instance of the collective. Given a random sequence of tosses with a fair die, the relative frequency of obtaining a two is equal to 1/6. The question raised by the problem of the single case is whether it is reasonable to ascribe the probability of 1/6, purportedly based on the relative frequency of the entire collective, to a single toss of the die. Obviously, unless the outcome of the tossing process is somehow determined – which it cannot be if the collective is truly random – then it is not rational to conclude that the relative frequency of the collective extends to the next toss of the die, or to any particular toss which is part of the collective. Thus it seems the only way a single case probability can be understood to be representative of the relative frequency of a given sequence, is if it is treated as a disguised statement about the long-run relative frequency of the collective.[35]

Popper's solution to the problem of single case probabilities is to argue that we can do without the idea of a collective to which the single case refers. In its place Popper proposes his own account of probability statements understood as measurements of physical propensities. However, it is not only the fact that the propensity theory can successfully deal with the problem of the single case that makes it relevant to the theory of probability. It also has the added quality of being able to explain how probability statements which apply to the single case, and are neither frequencies nor propensities to produce frequencies, can account for the statistical stability of random sequences.[36] Lastly, and most importantly, the propensity theory can account for the falsifiability of probability statements.

Popper's account of his propensity theory of probability begins with a distinction between absolute and relative probabilities. An absolute probability statement can be written as follows,

$$p(a) = r$$

and reads 'the absolute probability of *a is equal to r*'. A relative or conditional probability statement is written in the following way,

$$p(a, b) = r$$

and reads 'the relative probability of *a, given b, is equal to r*'.[37]

Popper's main interest is in the latter, especially since absolute probabilities can be defined in terms of relative probabilities.[38] And an emphasis on relative probabilities is more consistent with the situational, i.e., relational quality, of his general theory of propensity.[39] Also, working with relative probabilities is more conducive to the task of interpreting the more important experiments of quantum mechanics. For example, the Franck–Hertz experiment measures the relation between the rising voltage of electrons and a near discontinuous change in the interaction between electrons and gas atoms.[40] In fact, Popper's move from his own modified frequency theory to the propensity theory was brought about by his analysis of quantum theory. The propensity theory was initiated by the aim of providing an objective interpretation of the famous two-slit experiment of quantum physics.[41] According to Popper, the two-slit experiment reveals the reality of objective indeterminacies on the sub-atomic level. Consequently any sophisticated physical theory had to make a place for the role of objective probabilities in physics. The conclusion of his interpretation of the two-slit experiment is that probabilities are real physical propensities, and 'these propensities turn out to be *propensities to realise singular events*. It is this fact which led me [Popper] to reconsider the status of singular events within the frequency interpretation of probability.'[42]

Popper's account of his propensity interpretation of probability begins with a consideration of weighted possibilities. The classical or Laplacean account treats probabilities as the number of favorable possibilities divided by the total number of equal possibilities. Thus it is characterized by an insufficient generality, because it fails to consider possibilities that are unequal. Popper contends that a combination of the propensity theory with the measure-theoretic approach to probability statements both: (1) provides a sufficiently general account of probability, and more importantly (2) solves the problem of the single case. The solution of the latter problem directly underlies his account of the empirical refutability of probability statements.

According to Popper, consideration of weighted possibilities is necessary for all of science.[43] An example of a weighted possibility is a loaded die. In reference to the classical theory, a die that is loaded is characterized by an unequal possibility, and it cannot be an object of a probability statement. The aim of Popper's propensity theory is to demonstrate how it is possible to estimate the measure of a possibility, and why it is best to treat 'measures of possibilities as dispositions or tendencies or propensities'.[44]

Thus the propensity theory is also concerned with possibilities. But in contrast to the classical interpretation, the propensity theory can address both weighted possibilities and equal possibilities. Equal possibilities can be interpreted as weighted possibilities whose weights are equivalent.[45] Popper's propensity theory of probability is an attempt to show how weighted possibilities can be assigned numerical values. The method by which this end is attained is a statistical method. Given a sufficiently large repetition of events, we can statistically test or measure what the frequentist would call the 'relative frequency' of our test sequence. Popper states:

> Or, to be a little more explicit, the greater or smaller *frequency of occurrences* may be used as a test of whether a hypothetically attributed weight is indeed an adequate hypothesis.... So we use statistical averages in order to estimate the various weights of the various possibilities.[46]

And the relative frequency that is the object of our test is the result of the propensity inherent in the entire physical situation.[47] To make this point clear, let us examine more closely how the propensity theory handles the problem of the single case. As already stated, the frequency theory ascribes to a single event the relative frequency of the sequence to which it belongs. In contrast, Popper's propensity theory ascribes probability to a single event based upon it being a representative of a virtual or conceivable sequence. Thus if we express the probability of a single event as a relative probability statement, $p(a, b) = r$, then the difference between the frequency and the propensity interpretations can be found in how they interpret b. For the frequency theorist, the probability of the event a is conditional or relative to it being a member of b, that is, a member of the sequence having the relative frequency r. According to Popper, however, b represents the conditions inherent in the physical situation which 'produce the hidden propensity, and that give the single case a certain numerical probability'.[48]

Popper indicates the advantage of the propensity theory in regard to the problem of the single case through the following argument.[49] Consider a loaded die, whose probability of obtaining a six after many long sequences of throws is equal to 1/4. Next consider a sequence *b* consisting of: (1) a large number of throws with the loaded die, and (2) two or three throws with a fair die interpolated into the sequence of throws with the loaded die. Now Popper asserts that the probability of obtaining a six with the toss of the fair die in *b* is 1/6, regardless of: (1) the toss with the fair die being a member of the sequence having a relative frequency of 1/4, and (2) our having no knowledge of where in the sequence the toss with the fair die will occur. Now why is it that we ascribe the probability 1/6 to the toss of the fair die in *b*? The first aim of Popper's example is to indicate that we do not do so because the relative frequency of *b* has a bearing on the single case. According to Popper, we know that two or three throws with the fair die are not able to change the probability outcome of *b*. Thus the probability measurement for the fair die is not dependent upon it being a member of the sequence(s) of the loaded die. However, two or three tosses with a fair die are not enough for us to ascribe a probability of 1/6 to the throws with the fair die. (It is possible that in two tosses with a fair die a six-up may not appear at all.) Therefore, on Popper's reckoning, we have to postulate a virtual or conceivable sequence of tosses with a fair die. And we do this based on the conjecture that conditions exist that enable us in principle to generate the sequence required for the ascription of 1/6 to the two or three throws interpolated into the tosses with the loaded die. The generating conditions inherent in the physical situation make possible the probability 1/6 to be the probability outcome for both the single case and the sequence to which the single case belongs.

Now it becomes a testable fact of the physical world whether such a conceivable sequence can be sufficiently realized. A probabilistic hypothesis is tested by creating a particular experimental set-up and observing whether the sought-after single effect is attained if other conditions are varied.[50] Therefore probability statements are falsifiable given tests that allow for controls, that is, a test set-up, which is introduced into the objective physical situation. Further evidence of both (1) the importance of the general theory of propensity for physical theory and (2) the testability of physical propensities is apparent when we are forced to decide between an account of physical events in terms of propensities or an account in terms of deterministic forces.[51] Popper asks us to consider a test of a hypothesis positing an electrostatic force having a particular direction and intensity. A test can be carried out by placing a test body in the presence of the electrostatic force and observing whether a predicted direction and magnitude obtain. Now let us assume a sequence of tests that result in the direction of the accelerated body remaining constant, while the magnitude fluctuates. Furthermore, let us add that the test conditions have been kept as constant as possible. Now the determinist can attribute the fluctuations in magnitude to fluctuations in the initial conditions. However, Popper argues that if the conjecture of fluctuating initial conditions is untestable, then the determinist is left with the ad hoc conjecture of hidden, untestable, fluctuating initial conditions. Popper posits that a physical theory can be saved from such conjectures, only if it jettisons the idea of determinism and accepts an explanation of the above fluctuation in terms of propensities that are statistically testable in the manner explained above. Thus, in deciding whether to explain physical events in terms of either forces or propensities, questions of *testability* are decisive.[52]

Various criticisms have been lodged against Popper's theory, some placing greater emphasis on its logical/mathematical character and others critical of it as a physical theory. Briefly I consider each type of criticism in turn. Howson and Urbach, in their book *Scientific Reasoning*, present a criticism of Popper's propensity theory that posits his argument in support of objective single case probabilities involves an explicit contradiction.[53] At the outset let us assert that the criticism of Howson and Urbach is based on a fundamental misunderstanding of where Popper's argument in favor of the propensity theory stops. Nevertheless their presentation of it is correct up to a point. Having already rehearsed Popper's propensity theory in this chapter, there is no need to repeat it here. In accord with the above presentation of the propensity theory, Howson and Urbach correctly state that Popper's argument for the propensity theory (1) involves two throws with a fair die interpolated into a long sequence of tosses with a loaded die, and (2) concludes that probability outcomes are the result of the physical situation. Their criticism can be summarized as follows.

Let E_0 represent a toss of a fair die having specific physical parameters that regulate (determine) the toss of the die. Also, let the outcome of the toss of the die be other than six-up. Thus the relative frequency of obtaining a six-up in any sequence of repetitions of E_0 is 0.[54] Furthermore, consider experiment E that involves a sequence of tosses with a fair die, and has the relative frequency 1/6. Now Howson and Urbach conclude: 'The first throw of the fair die in Popper's imagined hybrid sequence [the interpolated sequence] is an instance of both E_0 and of E. Hence the single case probability of a six at the first throw of the die is both 0 and one sixth, and we have a contradiction.'[55]

This conclusion, however, is wrong because it merely establishes a point that Popper's argument made, and fails to see it as a stepping-stone to Popper's conclusion. Popper's aim upon considering the sequence characterized by the interpolation of the two fair tosses into the sequence of tosses with the loaded die is to establish this contradiction, and then to show that it could be resolved by introducing the idea of a virtual sequence. In his argument, Popper uses the example of the interpolated sequence to raise the question of why it is that even though we know that the two tosses with the fair die cannot produce a relative frequency of 1/6 (Howson and Urbach set it at 0), we also want to say that the probability outcome of a single toss with a die that we know to be fair is 1/6. Popper fully recognized this contradiction, and his answer to it is based on the following conjecture. Given that we are working with a fair die, we conjecture that if only the physical situation is such that we can construct a random sequence of tosses with the fair die (i.e., the virtual sequence is realized), then the relative frequency 1/6 obtains. Thus the relative frequency of the sequence characterized by tosses with the fair die is a function of the physical situation, and since the physical situation is basically the same for all tosses of the die, it follows that the relative frequency of the sequence can be ascribed to the single case. The contradiction is avoided if the world is such that the conceivable sequence is realizable. Moreover, the conjecture of the virtual sequence is not ad hoc because it is the object of a test.[56] Thus his argument asserts that the probability outcome of a single toss is a result of the underlying propensity inherent in the entire physical situation.

Concerning the status of the propensity theory as a physical theory O'Hear contrasts what he takes to be the predictability inherent in Newton's theory of forces and the purely conjectural quality characteristic of Popper's propensity theory:

> A difference between Popper and Newton is here that Newton's theory makes testable predictions. ... But we can guess at propensities only having observed how repeated instances of a given set-up behave. Popper's theory gives us no other way of postulating what propensities are obtaining.[57]

Two points need to be emphasized in response to O'Hear's criticism. First, the experimental set-up or the physical situation in general is structured in reference to background theories and initial conditions that don't preclude predictability. Second, it is surprising that having written about Popper's methodology of conjectures and refutations in previous chapters of his book that he disparages the conjectural status of Popper's theory of propensities. Admittedly, there is an important difference between a state of affairs that is purely conjectural (i.e., merely conceivable) and a state of affairs that is conjectural yet open to empirical tests, but having emphasized above that issues of testability are decisive to the propensity theory, there is no need to rehearse the point here.

To sum up, we can now see the importance of Popper's propensity theory for his account of objectivity. Because propensities have a mind-independent reality, objective science is the criticism of propensities, and truth is the correspondence between the statements of science and the relational dispositions that comprise the physical universe. This isn't to pretend that the debate over Popper's propensity theory is in any way settled; nevertheless, its ability to merit serious criticism is a sign of its fruitfulness.[58] For Popper, propensities are the fundamental structure by which the world shows itself to be recalcitrant to the theories we impose upon it. He concludes that a world fully amenable to critical inquiry is a world of propensities, for without a theory of propensities a great many scientific theories having a probabilistic form remain untestable. And in a more radical vein, human existence, the theories we construct, and the conditions for their criticism can be understood as propensities as well. In what follows, I argue Popper's treatment of human evolution and his World Three ontology are continuous with his propensity theory.

2. Evolutionary Epistemology

Many commentators have correctly pointed to the continuity between Popper's later evolutionary epistemology and his early account of scientific method.[59] The latter is characterized by the method of conjecture and refutation, and is ontologized in the former as the response of an organism to challenges proposed by the environment by means of trial and error gambits (conjectures) that are controlled by natural selection (refutation). Thus, Popper's account of propensities also underlies his evolutionary epistemology. His treatment of evolution emphasizes the importance of 'inborn expectations' and this phrase can be understood as another way of speaking about propensities that are to be realized in light of the demands of the physical situation, i.e., the environment. Thus Popper's theory of scientific method, his theory of propensities, and his Neo-Darwinism underwrite his explicit objectivist World Three ontology.

Briefly stated, Popper's evolution-based argument for objectivity asserts the following. Evolution of the physical universe has resulted in the emergence of human language as a problem-solving mechanism which contributes to the realization of an objective world through the formulation of problems, ideas, and mistakes whose objective status is evident because they are both (1) criticizable and (2) refer to an objective content which exceeds the informative content expressed in their original formulation, and which objectively affects the path of inquiry.

My analysis of both Popper's evolutionary epistemology and his objective ontology will treat the following areas of inquiry: (1) the logic of Neo-Darwinian evolutionary theory, (2) the parallels between the logic of evolution and critical rationalism, and (3) Popper's evolutionary ontology of World's One, Two, and Three and his argument for the objectivity of World Three.

2.1. Neo-Darwinian Evolution: A Standard Account

Popper finds the logic of evolution most amenable to his own falsificationist epistemology.[60] Fundamentally, Neo-Darwinian evolution is guided by two principal ideas.[61] The first idea is that all the various species descend from a single cladogenetic tree of life. Secondly, the species whose life histories are represented by this tree are the product of natural selection. Natural selection is characterized by three basic features: (1) variation grounded in genetic difference, (2) variation in fitness based upon genetic variation, and (3) the heritability of variations. Thus natural selection can be defined as 'heritable variation in fitness'.[62] Taking the aforementioned tree of life as the object of evolution, two types of evolution can be said to occur: (1) microevolution and (2) macroevolution. Microevolution concerns evolutionary changes of phenotypic and genotypic characteristics within a species. Macroevolution concerns the processes by which a new species comes into being. In general, 'evolution' denotes 'a process that converts variation among individuals *within an interbreeding group* into variation *between groups* in space and time'.[63] Granting a genetic basis to variation, the Neo-Darwinian account of evolution allows for evolution by means of either one, or by a combination of any of the following various mechanisms: natural selection, mutation, migration, random genetic drift, and sexual recombination. However, among these causes of evolution only natural selection 'can adapt populations to their environment'.[64]

Working from the above sketch, the logic of Neo-Darwinian evolution can be schematized as follows:

(1) Variation + (2) Inheritance + (3) Struggle for Survival + (4) Natural Selection, entail − New Species or Extinction

Parallels between Popper's early epistemology and Neo-Darwinian evolution are strongest at points (1) and (4) where variation is akin to novel scientific conjectures, and natural selection parallels critical refutations. But Popper argues for more than mere parallels. Popper claims that the logic of Neo-Darwinian evolution is the logic of falsification ontologized in actual organisms. The following is an examination of the interconnections

Cosmology and Propensities

between the logic of evolution and critical rationalism that are the basis for this claim. However, the harmony between methodology and ontology does not invite a naturalized epistemology. To think so is to give too much currency to the subject of epistemology as traditionally understood. Popper's philosophy from its inception replaces the issues of epistemology with issues of methodology: 'But what I call "methodology" should not be taken for an empirical science. I do not believe that it is possible to decide, by using the methods of an empirical science, such controversial questions as whether science actually uses a principle of induction or not.'[65]

2.2. The Logic of Evolution and Critical Rationalism

a. No Guarantee

Neo-Darwinian evolutionary theory guarantees neither that an organism's adaptations are an ideal fit with the environment, nor that an organism will survive confrontation with its environment. Analogously Popper's method of conjecture and refutation does not promise to certify the truth of a conjecture, or that a conjecture will survive criticism. It just so happens that both some conjectures and some organisms do survive criticism and selection by the environment respectively. In neither case does the fact of survival entail that they will survive in the future. The fact that each is corroborated by the fact of its present survival is not a license or an inference ticket to conclude that each will continue to survive. In each case, present survival is to be construed as a record of past performance.[66]

b. Growth by Criticism

According to Neo-Darwinian evolutionary theory, an organism insofar as it survives a threat to its existence from the environment may be said to incorporate the knowledge of that encounter into its survival mechanisms. Consequently, it can be said to grow in light of criticism from the environment. Similarly, Popper contends that his epistemology reveals how human knowledge grows under the selective pressure of experience. However, he identifies a distinction between a prescientific and a scientific stage of knowledge based upon the influence that the evolution of human language has upon the growth of knowledge.

c. Neo-Darwinism and Criticizability

Mathematical models more appropriately characterize Neo-Darwinian evolutionary theory than the traditional empirical laws that comprise much of the subject matter of physics and chemistry. Fisher's sex selection model and the Hardy–Wienberg Law, which are essential to contemporary evolutionary biology, are true independent of verifying or falsifying tests. For instance, the Hardy–Weinberg Law asserts that genotypic frequencies of sexually reproducing organisms will remain constant and are related to allele frequencies in the following way. Given the alleles A and a, the frequencies p and q of A and a respectively, can be ascertained by the following formula:

$$p^2_{(AA)} + 2pq_{(Aa)} + q^2_{(aa)} = 1 \quad ^{67}$$

Because such models are in principle non-empirical, predictions of empirical phenomena that could contradict them are impossible. But such models are able to convey a great deal about gene frequencies in sexually reproducing populations. Now the inclusion of such models as a part of science is in keeping with Popper's expanded notion of rational inquiry predicated on the distinction between the rationally criticizable and the non-criticizable. The above mathematical formula is criticizable as a purely formal representation of gene frequencies, e.g., whether it is mathematically consistent. But the question of whether it is applicable to any actual phenomena requires a biologist to go out and look at real populations. However, that it is applicable does not guarantee its truth because what it conveys about the distribution of alleles is mathematically true. Nevertheless, '[t]he Hardy–Weinberg formula is the starting theorem of evolutionary genetics'.[68] This follows because if there is significant deviation within genotypes from the predicted values specified by the above formula, then there is evidence of either: (1) non-random mating, (2) a change in allele frequency caused by evolution, or (3) both (1) and (2).[69] Thus, even though evolutionary biology incorporates non-empirical claims, it is in keeping with a Popperian account of science. This follows because, as I argued in the previous chapter, according to Popper non-empirical claims can have important implications for physical theory.

d. *Historicity*

Evolution is a historical process and so is characterized by historical hypotheses rather than nomological ones. And the growth of knowledge is also a historical process. It is a process that can be studied best by analyzing how it is that our theories serve as tentative solutions to the problem situations that gave rise to them. Thus, the history of the growth of knowledge – which can be studied best by studying the growth of scientific knowledge – has a biological analogue. Life proceeds like scientific discovery.[70]

e. Organisms are Problem Solvers

For Popper organisms are problem solvers. Essentially the problem that organisms are attempting to solve is how to avoid their own extinction, that is, criticism that will result in their death. The problems that constitute the physical situation within which an organism must struggle are, in part, the result of its own genetic makeup. A frog, for example, only acknowledges a fly as a food source when the fly is in motion. Thus the frog's adaptation to the environment, i.e., physical situation, is inadequate and this inadequacy forms part of its problem situation.[71] Concomitantly, the fact that an organism's environment may introduce novelties to which it is ill adapted comprises the other feature of its physical situation. Now Popper contends that an organism's adaptations are the biological analogues of scientific theories. In fact, even an organism's organs, for example its eyes, can be said to be theory-laden. On the purely biological level such theories are an organism's inborn expectations that inform its approach to the environment. Accordingly these theories are a type of conjectural knowledge that an organism has prior to its confrontation with its physical situation. And this fact indicates 'induction breaks down before taking its first step'.[72] If inductive inference is to be understood as the building of theories from sense data, and if Neo-Darwinian evolutionary theory is retainable as true, then induction and its consequent psychologism are false, because Neo-Darwinian theory claims that 'there are no such things as sense data or perceptions which are not built upon theories (or expectations – that is, the biological predecessors of linguistically formulated theories)'.[73]

f. Genetic Basis

Furthermore, as argued in Chapter 1 above, the fact of inborn expectations or theories induces Popper, in a manner reminiscent of Kant's epistemology, to describe organisms as possessing both a priori and a posteriori knowledge. Popper is not asserting that the inborn expectations which correspond to a priori knowledge are necessarily true, i.e., a perfect fit with the environment. In fact, he explicitly denies this because an organism's expectations, or theories, or conjectural knowledge claims often fail to ensure its survival. Like Kant, however, Popper posits that there is no such thing as uninterpreted data.[74]

g. Situational Logic

It is manifest given his account of propensities that an organism's inborn expectations coupled with the changing features of its environment can be understood to constitute a propensity for a particular experience in the life history of the organism. This fact is generalized to explain Neo-Darwinian evolution as constituting what Popper calls 'situational logic'.[75] On Popper's reckoning, the facts of adaptive variability and environmental instability comprise a physical situation that makes Neo-Darwinian evolution 'almost logically necessary' (ibid. 134). This follows, according to Popper, because without a natural mechanism whose logical form parallels the logic of falsification both survival and the growth of knowledge appear impossible. Survival and the growth of knowledge both require a means of eliminating error.

Given his account of the relationship between organisms and their environment Popper asserts that the following schema captures the situational logic of Neo-Darwinian evolution. Let P, TS, and EE designate: (1) problem situation, (2) tentative solution (Popper sometimes labels this step in the process TT, for tentative theory), and (3) error elimination, respectively.

$$TS_1$$
$$P_1 - TS_2 - EE - P_2$$
$$TS_n$$

h. Background Knowledge

Upon relating this schema to Popper's account of scientific method and the growth of knowledge we can see that organisms, like scientific theories, confront problem situations that are formed in reference to a particular background.[76] In Popper's accounts of both scientific method and Neo-Darwinian evolutionary theory problems give rise to theories and adaptations, respectively.

Our understanding of the relationship between the logic of falsification and Neo-Darwinian theory on the issue of the growth of knowledge can be better illuminated by treating the following analogy – Falsification : Induction :: Neo-Darwinism : Lamarckism. Both induction and Lamarckism attempt to establish respectively a hypothesis or the adaptation of an organism based upon repeated experiences. However, if a theory or an adaptation can be explained by two distinct kinds of repeated experiences, then neither Lamarckism nor the method of induction can adjudicate between the competing explanations. In contrast, Popper asserts that both falsificationism and Neo-Darwinian evolution have a logical form such that a *preference* between respective competing explanations of both a particular theory and an adaptation can be established. It was this feature of both falsificationism and Neo-Darwinism that contributed to Popper's identification of the two.[77] Indeed Popper asserts the identity of the two in no uncertain terms:

> This statement of the situation is meant to describe how knowledge really grows. It is not meant metaphorically, though of course it makes use of metaphors. The theory of knowledge which I propose is a largely Darwinian theory of the growth of knowledge. From the amoeba to Einstein, the growth of knowledge is always the same: we try to solve our problems, and to obtain, by a process of elimination, something approaching adequacy in our tentative solutions.[78]

3. The Fundamental Difference: Towards an Evolutionary Ontology

The fundamental difference between Popper's critical rationalism and Neo-Darwinian evolutionary theory is not a feature of their logical structures. Each is characterized by a logical form that makes possible both criticism and, respectively, a preference between scientific theories and a preference between adaptations which are the biological analogues of theories. Where the two differ is in their approach to feedback in the form of criticism. The scientist in employing the method of falsification seeks out criticism of his theory; but an organism avoids criticism at all cost. We can see the reason for this difference if we follow up Popper's comparison of Einstein and an amoeba. The latter perishes when its inborn expectations are falsified; the former grows in knowledge when his theories are criticized because he learns to replace his tentative solution to a problem with a new solution, and if necessary to replace the problem itself. Thus the difference between Einstein and the amoeba is that the former develops a critical attitude that allows his theories to perish in his stead. This critical attitude is the basis for the distinction between a scientific and a pre-scientific stage of knowledge. Scientific knowledge is characterized by the criticism of scientific theories, and such criticism is made possible by the formulation of theories in language which in turn renders them public objects, and hence subject to criticism in the form of intersubjective tests. Thus a critical attitude, which differentiates the pre-scientific stage of knowledge from the scientific stage, is a result of the evolution of language. However, it is important to remember that both kinds of knowledge share the same logical form, and that it is how criticism is approached at each stage that differentiates them.

At the pre-scientific stage criticism is to be avoided, and dogmatism is optimal. In the case of scientific knowledge, Popper asserts that criticism is essential to growth. And because the scientist can be said to learn from his mistakes (i.e., receive instruction from the environment) Popper argues that falsificationism and Neo-Darwinian evolution simulate induction and Lamarckism.[79] Thus Lamarckism and induction are respectively first approximations to Neo-Darwinian evolutionary theory and falsification. Moreover, both Lamarckism and Neo-Darwinism attempt to account for why given a particular physical situation certain mutations within a species have survival value. Thus both attempt to explain the appearance of purpose in the world. Indeed, for Popper, the great explanatory power of Neo-Darwinian evolutionary theory is evidenced by how easily it can treat of this issue. Popper asserts:

> For Darwin's theory of natural selection showed that it is *in principle possible to reduce teleology to causation by explaining, in purely physical terms, the existence of design and purpose in the world*. What Darwin showed us was that the mechanism of natural selection can, in principle, simulate the actions of the Creator, and His purpose and design, and that it can also simulate rational human action directed towards a purpose or aim.[80]

Thus, according to Popper, Neo-Darwinian evolution is characterized by a system of 'plastic controls'.[81] Higher-level adaptations direct and control the functioning of lower-level adaptations, but allow for feedback . This point is important for the possibility of criticism, i.e., error elimination, and therefore it is important to our general understanding of the logic of evolution. If knowledge truly grows through criticism, then our theories cannot be reduced to 'superfluous by-products (epiphenomena) of physical events'.[82] Without a degree of 'plasticity', both theory and observation are equivalent to predetermined physical events. Consequently, the growth of knowledge through criticism requires a balance to be struck between determinism and indeterminism. Neo-Darwinian evolutionary theory provides such a balance because trial and error gambits, or variations, are random and hence indeterminate; but evolution is also determinate because the laws of heredity are fixed. This interplay between indeterminate and determinate factors is also reflected in how higher-level adaptations exercise control over lower-level adaptations yet allow for novelty. Popper argues that in humans there is both a higher language adaptation ('human language') and a lower language adaptation ('animal language'); however, human language has developed beyond animal language. Animal language is characterized by two functions.[83] The first is an expressive function that indicates the state of an organism via a linguistic sign – e.g., a smile. Second, there is a reaction to this sign by another organism resulting in a signal being established between the two. Popper calls this function the signaling function. Now Popper argues that human language has developed beyond animal language, yet builds upon it resulting in two functions fundamental to critical inquiry, and to his account of objectivity as well. These two functions are the descriptive and the argumentative functions of human language.[84] Descriptive statements that can be classified as either true or false characterize the descriptive function. The truth or falsity of a descriptive statement is ascertained by the argumentative function, which employs the method of conjecture and refutation. Given these higher-level functions, Popper claims that a discussion at a scientific conference, for example, is evidence of the 'plasticity' of language as an adaptation.[85] A conference may give rise to expressions of being exciting and enjoyable, and these expressions may signal similar symptoms to other participants. However,

> there is no doubt that up to a point these symptoms and releasing signals will be due to and controlled by the scientific content of the discussion; and since this will be of a *descriptive and of an argumentative nature* the lower functions will be controlled by the higher ones.[86]

Popper's position on the issue of the interdependence between the higher and lower adaptations is that human language is irreducible to animal language but has emerged from it via evolution. Similarly, the higher language functions are not reducible to the lower functions. Moreover, Popper now asserts that understanding the continuity between animal and human knowledge is 'the main task of the theory of human knowledge'.[87] Central to such a project is to understand how human language has made possible the critical activity that Popper understands to differentiate Einstein from the amoeba while at the same time recognizing their interconnectedness as problem solvers.

Treating language as a self-contained subject matter, philosophers have mostly addressed the problem of language independent of what science can tell us. This isn't, however, to deny the study of scientific semantics as pursued by Tarski and like-minded thinkers. Nevertheless, to a large extent philosophers have formulated and continue to formulate problems about language without understanding how language as a problem for science should inform their own inquiries. Evolutionary theory invites a more expansive approach. Consider how biological insights change the theory of meaning. An evolutionary approach to language of the kind advocated by Popper gives the meaning-as-representation and meaning-as-use debate a whole new orientation. Look at language from the perspective of homology. 'Homology' denotes 'the correspondence between two structures due to inheritance from a common ancestor'. For example, human eyes and frog eyes have the same use but represent the world differently. Language understood as a biological adaptation never allows issues of meaning to be separated from issues of use, but at the same time some representations are more useful than others, more adaptive to the environment, and thus a better fit; therefore, the distinction between meaning-as-use and meaning-as-representation is worth keeping without raising issues of which is more foundational or has logical priority. Consider another example. Russell's theory of definite descriptions as part of the meaning-as-representation approach to language is an attempt to explain the denotation of empty expressions. From the perspective of biology, and thereby viewing language as an adaptation, there is in the background of the tradition that Russell is responding to the following assumption: every 'meaningful' adaptation must reference something. However, Neo-Darwinian evolutionary theory asks us to consider that there are variations that do not correspond to any state of affairs and thus can't be said to represent anything. Ultimately, Russell's problem situation becomes far less troubling in light of the evolutionary approach to problem solving that Popper advocates. Consequently, evolutionary theory should change the way philosophers approach issues of language. For Popper it definitely changed his approach to ontology. Via human language evolution has made possible a realm of statements-in-themselves that accounts for knowledge without a knowing subject.

3.1. An Evolutionary Ontology

In this section, our aim is to treat Popper's evolutionary ontology. Popper's well-known pluralistic ontology contains his most explicit claim for an objective, mind-independent world. His ontology is characterized by three worlds: World One, World Two, and World Three. Briefly, World One is the world of physical objects; World Two comprises states of consciousness; and World Three contains the logical contents of thought.[88] These three worlds are the product of evolution. They interact with one another at different levels, and this interaction is both structured by evolutionary developments and contributes to the growth of knowledge. Of these three worlds, World Three is the locus of objectivity, and hence it is the focus of this section. But before analyzing Popper's account of the three worlds in greater detail let us first address the question of why Popper believes that more than the first two worlds are necessary to any ontology mindful of the conclusions of contemporary science, that is, evolutionary theory.

Popper is aware that he is not the first philosopher to assert the reality of an objective third world. Three 'realms' or worlds, for example, characterize Plato's metaphysics, and one of these three realms is the realm of the 'Forms' which is completely mind-independent.[89] But unlike Plato's ontology, Popper's objective realm is, in part, the result of rational scientific inquiry and not simply a prerequisite for the possibility of science. Popper believes philosophical speculation concerning reality shouldn't be conducted independent of mature scientific inquiry. Accordingly, his three-world ontology can be understood as satisfying the demands of Neo-Darwinian evolutionary theory.

For Popper, Neo-Darwinian evolutionary theory evidences the theory-laden nature of observations. But at the same time the dualism of theory and observation is an important explanatory framework for evolutionary biology.[90] The dualism between theory and observation corresponds in organisms to the distinction between an adaptive expectation (theory) and sense data. This dualism which evolutionary biology claims characterizes the life of organisms, invites the introduction of a real-world construct according to which a scientist can assess whether data and expectation fit. Without an assessment of fitness in reference to a real world, the distinction between theory (adaptation) and observation (sense data) collapses. Popper captures this insight when he states, 'the amoeba cannot be critical *vis-à-vis* its expectations or hypotheses [i.e., theories]; it cannot be critical because it cannot *face* its hypotheses: they are part of it'.[91] Thus the conjecture of a real physical world is added to the above dualism, because it is retainable in light of actual scientific inquiry, that is, evolutionary biology, which retains along with the dualism of theory and data the idea of a mind-independent world. Campbell nicely captures the meaning of Popper's treatment when he writes:

> Biological theories of evolution, whether Lamarckian or Darwinian, are profoundly committed to an organism–environment dualism, which when extended into the evolution of sense organ, perceptual and learning functions, becomes a dualism of an organism's knowledge of the environment versus the environment itself. An evolutionary epistemologist is at this level doing "epistemology of the other one," studying the relationship of an animals cognitive capacities and the environment they are designed to cognize, both of which the epistemologist knows only in the hypothetical-contingent manner of science. ... At this level he has no hesitancy to include a "real world" concept, even though he may recognize that his own knowledge of the world even with instrumental augmentation is partial and limited in ways analogous to the limitations of the animal whose

epistemology he studies. Having thus made the real-world assumption in this part of his evolutionary epistemology, he is not adding an unneeded assumption when he assumes the same predicament for man and science as knowers.[92]

Thus Neo-Darwinian evolutionary theory demands a World Three ontology. Unfortunately, Popper's defense of a mind-independent world does not at all emphasize the hypothetical/conjectural realism that evolutionary theory makes available.[93] However, it is important to mention, if it means turning to the topic of language only briefly, that in contrast to Davidson, Popper's evolutionism commits him to the notion of 'conceptual schemes' – what Davidson calls the third dogma of empiricism – since conceptual schemes are requisite to assess theory/adaptation critically and fit with the environment. Here again is an instance where methodological concerns within science trump purely philosophical speculations.[94]

a. World Three and Its Interactions

World Three, according to Popper, is the product of the evolution of human language. Human language is characterized by its descriptive and argumentative functions, and because of these there exists theories, mistakes, errors, problems and problem situations, and descriptive statements and rational arguments. Also, other entities are the tools, the works of art, and the cultural and social institutions that are made possible by the objective content of World Three.[95] Popper argues that World Three entities originate in World Two, but are not reducible to it. Thus World Three is partially autonomous of World Two.

Popper's distinction between these two worlds parallels the distinction he made in *The Logic of Scientific Discovery* between the psychology of knowledge and the logic of knowledge.[96] There, Popper argues that the logical analysis of a theory is distinct from and not reducible to how a person conceives or invents a scientific conjecture. Analogously, the subjective thought processes of World Two differ from the objective or logical content of the theories that are the result of World Two activity. Popper points to the similarity between his own views and those expressed by Frege. He quotes Frege as follows: ' "I understand by a *thought* not the subjective act of thinking but its *objective content*." '[97] World Three comprises the objective content of thought which can be analyzed in terms of logical categories such as: incompatibility, equivalence, deducibility, and contradiction.[98]

For Popper, epistemology is the study of how scientific knowledge grows. However, he thinks philosophers have traditionally occupied themselves with knowledge in the subjective sense of World Two. Consequently, they concerned themselves with such issues as the origin and foundation of beliefs, and an analysis of the psychological mechanisms that contribute to belief formation. Hume's epistemology would be a classic example of a subjective epistemology that 'has studied knowledge or thought in a subjective sense – in the sense of the ordinary usage of the words "I know" or "I am thinking" '.[99] And on Popper's reckoning, all such epistemologies are irrelevant because scientific knowledge is not knowledge in the subjective sense of 'I know', but concerns the objective content of theories.[100]

Popper establishes the priority of World Three as the object of scientific knowledge upon the fact that scientists both construct and critically test theories whose objective content is about World Two activity (e.g. cognitive psychology). His point is that although information about the subjective thought processes of a scientist may indicate how she arrived at a particular theory; this information provides no clue as to the truth of the theory. Thus it is the objective content of World Three that is the object of scientific inquiry. For Popper, scientists analyze the content of theories and seek to identify false consequences of them in order to classify them as false, or true if they withstand such attempts at falsification.

Furthermore, the above argument for World Three as the object of science manifests the nature of World Two and World Three interactions. World Two gives rise to World Three through the formulation of problems and arguments, and the objective content of these problem situations and arguments have a feedback effect on World Two because the content of our ideas often affects how we relate subjective experiences. Consequently, human language, essentially in its descriptive and argumentative dimensions, plays a central role in Popper's evolutionary ontology. In fact, human language belongs to all three worlds.[101] It begins in World Two with subjective thought processes, but takes on a partial autonomy made possible by its World Three content, which in turn is instantiated in World One objects such as books, computers, and works of art. And by being studied, the World One objects can lead to the formulation of new ideas in the minds of individuals and thus affect World Three. Thus interaction between World Three and World One is mediated by World Two, which is itself the result of the evolution of language.

b. *The Objectivity of World Three*

In what follows, I contend Popper's arguments for the objectivity of World Three are best understood as arguments for particular kinds of objective propensities. Such a claim emphasizes the continuity of his early and later thought, and that his case for the reality of World Three is supported by his general theory of propensity. This means his objectivist ontology is commensurate with his argument for objective propensities developed in conjunction with his work on probability. In his treatment of probability theory Popper argues that probability statements and the propensities that are the basis for them are not reducible to subjective states of ignorance. But this is not to say, however, that our subjective states cannot be characterized by propensities of their own. Thus what is at issue is whether World Three entities are unique propensities not reducible to World Two or, for that matter, to World One.

In the first chapter of *A World of Propensities*, subtitled 'Two New Views of Causality', Popper's aim, in part, is to invoke his treatment of propensities as a panacea to both determinism in physics and 'the ideology of determinism in human concerns'.[102] His treatment of the latter kind of determinism, that is, the problem of free will, leads him to characterize World Three entities as propensities. If the content of our theories and guesses were strictly determined, then the growth of knowledge through criticism would be impossible. The content of our theories makes possible the growth of knowledge by creating a situation or context in which criticism is possible. For Popper a propensity is characterized by more than mere possibilities. Propensities are possibilities that are weighted in light of an overall situation. In the case of World Three propensities the situation is the content and critical milieu surrounding a theory. Moreover, like physical propensities our theories are causally efficacious, for example, by inspiring technological achievements. Additionally, they have a history, and like physical propensities a temporal quality. Popper's world of propensities incorporates the denizens of World Three. In *A World of Propensities* he makes the transition from a discussion of physical propensities to World Three propensities in the following text:

> Now, in our real changing world, the situation and, with it, the possibilities, and thus the propensities, change all the time. They certainly may change if we, or any other organisms, *prefer* one possibility to another; or if we *discover* a possibility where we have not seen one before. Our very understanding of the world changes the conditions of the changing world; and so do our wishes, our preferences, our motivations, our hopes, our dreams, our phantasies, our hypotheses, our theories. Even our erroneous theories change the world, although our correct theories may, as a rule, have a more lasting influence.[103]

Obviously, in referring to the influence or efficacy of our theories Popper has in mind their objective content, that is, their status as denizens of World

Three. It is only if the denizens of World Three are understood as propensities and thus as objectively indeterminate, that human creativity is possible. Moreover, in the background of Popper's ontology is the theory of evolution. The three worlds depict the course of evolutionary development; World One makes possible World Two consciousness, which in turn produces the partially autonomous World Three, and Worlds Two and Three together contribute to the realization of World one artifacts. Each world is to be understood as an emergent propensity constituted both by the situational logic of one or more of the other worlds and its own structural properties. And this situational logic itself constitutes a kind of propensity – a propensity for an emergent world. Thus the fundamental question underlying the issue of World Three autonomy is whether genuine emergence is possible, that is, the emergence of completely new propensities. However, Popper doesn't address this issue head on; instead, he takes the approach that if he can show that World Three isn't reducible to the other two worlds, then he has successfully made his case for the autonomy of World Three. This is unfortunate because Popper could have brought his impressive logical skills to address the very technical problems of emergence and supervenience to be found in the philosophical literature.

The principal opponents of Popper's arguments for the autonomy of an emergent World Three are the subjectivists and materialists who would reduce World Three to Worlds Two and One, respectively. Popper's development of his triadic ontology was initiated by the intrusion of subjectivism into quantum theory.[104] Thus the subjectivist challenge that aims to reduce World Three to World Two is of special interest to him. Popper proffers several specific arguments for the objective reality of World Three.[105] These arguments can be divided according to the two kinds of objective propensities they make known: (1) propensities concurrent with language, and (2) propensities associated with both the products of language and with World One objects. Given this division let us consider Popper's arguments for each.

c. Objective Propensities Concurrent with Language

According to Popper, it is only when our ideas are formulated in language that they attain an objective status. In part, this is because the formulation of a subjective experience in language renders it a public object open to criticism via intersubjective tests.[106] Thus, in a manner of speaking, by putting a subjective experience into language a person gives it away, and so renders it less subjective because it is open to criticisms and interpretations not of her making. One may want to reply, however, that the fact that we attain a language for our subjective experiences contributes nothing to an argument for the objectivity of World Three, because we often use language to formulate and criticize our ideas subjectively before rendering them public objects through communication to others. Simply put, we think to ourselves. However, even at the level of interior thought-speech there exists an inherent propensity to intersubjective criticism. And this propensity to be criticized is an objective feature of the logical content of an idea once it is formulated in language, regardless of whether it is ever criticized or communicated. Let us call this objective propensity the objective critical propensity of language.

Also, as previously noted, Popper argues as part of his treatment of propensities that the scientific determinism of the Laplacean superman (the classic example of radical subjectivism) is refuted by the special theory of relativity. A subjectivist understands Laplace's superman argument to reduce the probability of all events to the state of knowledge of his superman. Now the subjectivist must explain how it is that an argument for the reduction of all phenomena to World Two is refuted by the special theory of relativity, which according to the subjectivist is a World Two construct? A possible answer is that there is nothing wrong with World Two containing contradictory statements or beliefs. However, it is obvious that such statements are not contradictory as instances of mental states or as subjective experiences per se. Thus they must be contradictory in terms of their objective content. Accordingly, concurrent with the formulation of a subjective experience in language there arises an objective propensity to criticism.[107]

d. Propensities and Products

Popper argues that once a problem or argument is formulated in language it has objective unforeseen consequences, i.e., objective products. One of his favorite examples to support this claim is the creation of the sequence of natural numbers. Popper claims that the existence of this sequence has led to the unforeseen and unintended problem of whether the sequence of prime numbers is infinite.[108] For Popper this problem existed to be discovered and is not the creative work of any individual or group. Thus Popper asserts:

> Knowledge in this objective sense is totally independent of anybody's claim to know; it is also independent of anybody's belief, or disposition to assent; or to assert, or to act. Knowledge in the objective sense is *knowledge without a knower*: it is *knowledge without a knowing subject*.[109]

Popper further supports this point through the following example involving a World One object. He asks us to consider a World One object such as a book of logarithm tables.[110] Such a book contains many statements with a content that may both never have been known, nor will be known by any individual (the book can contain logarithms to the fiftieth decimal which may have been carried out by a computer). Thus Popper believes himself to be justified when he asserts that 'each of these figures contains what I call "objective knowledge"; and the question of whether or not I am entitled to call it by this name is of no interest'.[111] Thus, on his reckoning, knowledge exists without a knower. Moreover, this example seems to imply that what makes the ink marks in the book of logarithms knowledge is that they have a propensity to be known. Thus Third World entities can be understood as the dispositional properties of World One objects and so bypass interacting on World Two level. Popper states:

> [that] which turns black ink spots on white paper into a book, or an instance of knowledge in the objective sense ... is its possibility or potentiality of being understood, its dispositional character of being understood or interpreted, or misunderstood or misinterpreted, which makes a thing a book. And this potentiality or disposition may exist without ever being actualized or realized.[112]

Furthermore, Popper claims that his account of World Three objective entities can be further strengthened by showing that the philosophical arguments to support their autonomy are also supported by biological analogues. First, as we stated above, Popper claims that the problem situation of an organism is initiated more by its preferences than by any immediate threat to its survival. Preferences function as a kind of internal selection pressure; that is, they originate within the organism and direct it to overcome particular challenges that may frustrate the satisfaction of the preference. In fact, Popper conjectures that there exists a type of gene controlling behavior, called by him a b-gene, which can be further subdivided into p-genes (preference controllers) and s-genes (skill controllers).[113] And the preferences which result from these genes are to be understood as 'being dispositions, [which] are not so very far removed from propensities'.[114] Now these preferences in being a type of adaptation are products of the evolutionary process akin to theories, and so like theories are characterized by an objective content.

Second, and more importantly, knowledge in its broadest sense can be taken to include the objective products of evolution. Examples of such products are, of course, the scientific theories of humans, and also the nests of birds and other animals.[115] All such products can be studied objectively, that is, we can analyze them not in terms of the subjective thought processes that underlie their production, but instead as products developed in response to a particular problem situation. Indeed, Popper calls the method of approaching problems from an analysis of the products created to solve them 'the "objective" approach or the "third-world" approach'. The approach is one of working backward from effects to causes. What makes this approach objective is not simply that it is unconcerned with uncovering the psychological or sociological processes that contributed to the creation of a product. More importantly, for Popper, an analysis of the products of evolution without regard for the subjective thought processes that brought them about, reveals that both the problem situation and the product intended to solve it are characterized by propensities. Consider a bird's nest: whether it was built by the bird or provided by humans, it exists to solve the problem of establishing an adequate ecological niche.[116] The problem of whether or not the nest is adequate is an objective problem independent, for example, of the intentions that humans may have had in providing it. Moreover, if a specific bird or group of birds abandons the nest it has the propensity to be inhabited by other organisms. On Popper's reckoning, an account of the subjective thought processes that led to the creation of the bird's nest, e.g., a desire to engage in bird watching, fails to characterize scientific inquiry as it investigates the objective features of the product.

The 'approach' from the side of products, especially the analysis of them as World One objects, is further evidence of the reality of World Three entities. According to Popper, technology is an excellent example of how the objective content of World Three is instantiated in World One. This instantiation is evidence that World Three entities are 'real'. In one of his few discussions of the meaning of a term Popper states:

> I propose to say that something exists, or is real, if and only if it can be kicked and can, in principle, kick back; to put it a little more generally, I propose to say that something exists, or is real, if and only if it can *interact* with members of World 1, with hard, physical, bodies.[117]

And Popper takes it as obvious that scientific theories are exploited by technology.

Additionally, it is both interesting and important that Popper's handling of the term 'real' squares with our account of World Three entities as propensities. In the above quotation Popper identifies the real with that which "can, in principle" affect World One and be affected by it. Now the phrase "in principle" points to the real itself as having the quality of a propensity, that is, a dispositional structure. However, Popper's World Three entities are not real in the same way that Plato's Forms are real. According to Plato, the Forms are unchanging, atemporal, and true. In contrast, Popper's World Three has a history (it only came into being with the evolution of human language), it contains false statements, and as a result of increased World Two activity it constantly changes because of the addition of new objective content which can stand in novel logical relationships to already existing World Three denizens. Moreover, Popper points out that for Plato argumentation was the means by which the Forms were known, whereas in his own ontology arguments are one of the most important World Three entities. Ultimately, World Three entities are primarily to be understood as objective propensities. They are characterized by: (1) an initial propensity to criticism concomitant with their formulation in language, (2) a propensity to engender the discovery of other problems given the current situational logic of World Three, and (3) can be understood as the dispositional properties of World One objects.

Despite Popper's worthwhile intentions of combating subjectivism, his arguments for World Three as concurrent with language seem woefully inadequate and ignorant of developments in the philosophy of language. To briefly consider two of Popper's contemporaries, Wittgenstein and Quine, it is amazing that Popper is not aware of the anti-essentialist elements in the later-Wittgenstein's writings and in Quine's 'Two Dogmas'[118] as directed at his claim for an immaterial proposition or meaning underlying a linguistic expression. Quine's rejection of meaning as cognitive synonymy and his adamant extensionalist approach to language is an important foil to Popper's World Three ontology. Quine's argument is not concerned with how meanings or propositional content relate to the user of a language, and thus his approach is very Popperian, because Popper's rejection of justificationism in epistemological matters consistently disavows as important how evidence relates to the beliefs or confidence of the knower. Popper offers his readers many examples of what he intends by the objective content of language but concerning this issue, and contrary to practice, he needs an argument from an analysis of language explaining how language has an immaterial objective content. Therefore, his defense of World Three from the perspective of World One artifacts is the more promising of his two approaches, because it opens up the possibility of studying language as an exosomatic byproduct of evolution and thus as an object of inquiry consistent with the conjectural realism that evolutionary theory and his own methodology seems to invite.

Notes

1. Popper (1990) 9.
2. Ibid. 12.

3. See Popper (1983) 359 and (1959a) 37, where Popper has compared his propensity theory to Aristotle's account of potentialities. See Popper (1983) 282, footnote 2, concerning others who have compared it to C. S. Pierce's treatment of 'habit' or 'would-be'. Any strict identity between these theories and Popper's is incorrect because Popper's theory characterizes propensities as relational. In contrast, the other theories treat their respective accounts of potentialities and habits as properties of things.
4. Popper (1990) 13.
5. Ibid.
6. Simkin (1993) 66.
7. Popper argues that in science we often explain the known by the unknown, that is, scientific test are often directed at falsifying conjectures concerning structural properties which are not directly testable. Popper (1983) 162, 192 and (1963) 174.
8. Popper and Eccles (1977) 25.
9. Ibid.
10. Popper (1982b) 159.
11. Popper (1990) 14.
12. Popper (1959b) 30.
13. Suppes (1974) 760–74.
14. Popper *et al.* (1974) 1130.
15. Popper (1959a) 146.
16. Ibid. 190.
17. Ibid. 146.
18. Ibid. 151.
19. Consistent with his emphasis on the triviality of definitions, Popper does not believe that there is any one definition of probability. According to Popper, the proper approach to mathematical probability theory is to: (1) develop a consistent probability calculus, and (2) interpret it in a way that satisfies the demands of physical theory. Popper *et al.* (1974) 1126ff.
20. Popper (1959a) 150.
21. A treatment of the Laplacean account will be reserved until we treat Popper's propensity theory of probability.
22. Popper (1959a) 148.
23. Ibid. 149.
24. Ibid.
25. Popper (1983) 391.
26. Von Mises (1957).
27. Popper (1959a) 151–2.
28. Howson and Urbach (1989) 208.
29. Popper (1959a) 153.
30. Ibid. 360.
31. Ibid.

32. Popper (1959a) 292, footnote 1, argues that according to von Mises' frequency theory, a short, regular commencing sequence such as – 00 II 00 II – cannot be understood as random without the added idea of deferred randomness. Popper asserts that random sequences of any length can be constructed without the presupposition of deferred randomness, along the following lines. Let $x = n + 1$. Construct a table of all the 2^x possible x-tuples of ones and zeros. Construct a sequence by tabulating the last of the x-tuples, consisting of x ones, and check it off the table. Proceed according to the following rules: (1) always add a zero to the segment if permissible ('permissible' denotes that the last x-tuple created has not yet occurred and has not been checked off the table), and (2) if this is not permissible add a one. Continue in this manner until all the x-tuples have been checked off. The result is a sequence of the length $2^x + x - 1$. This sequence has: (1) a generating period of $2^x = 2^{n+1}$ of an n-free alternative, and (2) added to it the first n elements of the next period. Thus Popper claims to have constructed 'a sequence whose degree of n randomness (that is, its n-freedom from after-effects) grows with the length of the sequence as quickly as is mathematically possible' (ibid. 360).
33. O'Hear (1980) 130.
34. Miller (1994) 180.
35. Kahane (1990) 364.
36. Miller (1994) 179.
37. Popper (1983) 283–4.
38. See Popper's reply to Suppes' article in Popper (1974) 1132.
39. Popper (1990) 14.
40. Ibid. 15.
41. Popper (1959b) 28.
42. Ibid. For Popper, the propensities referred to are relational propensities in the sense discussed in section A, above. It is the entire physical situation that causes the objective indeterminacy observed in the two-slit experiment. The path a sub-atomic particle takes in the two-slit experiment is not a property of the sub-atomic particle per se. Thus Popper's reading of the two-slit experiment in this way lends further support to our treatment of his account of physical propensities.
43. Popper (1990) 10.
44. Popper (1983a) 358.
45. Popper (1990) 10.
46. Ibid. 11.
47. Popper's earliest accounts of the propensity theory placed special emphasis on the experimental set-up. But eventually he came to emphasize the entire physical situation, because emphasizing the former seemed to have a subjective flavor. Compare (1959a) 38 and (1990) 12. Also, although he argues for the superiority of the propensity theory, nevertheless, it is the case that the frequency theory contributes significantly to any tenable account of probability. For within Popper's propensity theory, probability statements expressing relative frequencies are the object of tests. See Popper (1983a) 361 and (1990) 11. Moreover, the conjoining of his

propensity theory with the measure-theoretic approach to probability statements lends the frequency theory 'a kind of *post mortem* justification; for the frequency theory becomes "almost deducible" from the measure-theoretical approach'. Popper (1983a) 347.
48. Popper (1983a) 287.
49. Popper (1959a) 31–7 and (1983a) 350–9.
50. Popper (1983) 289.
51. Popper (1982a) 93–5.
52. Ibid. 94. Italics are mine.
53. Howson and Urbach (1989) 222–5.
54. Ibid. 223.
55. Ibid.
56. Indeed, Howson and Urbach (1989) invokes a similar use of 'unimpeded conjecture' in response to Popper's criticism of the subjective interpretation of probability. See *op. cit.* 259–62.
57. O' Hear (1980) 133.
58. Keuth (2005) 189–90.
59. Magee (1985) and Watkins (1974) 371–412.
60. Initially Popper did not consider evolutionary theory to be an actual scientific theory and instead identified it as a 'metaphysical research program'. See Popper (1974) 151, 167ff. Consequently, his treatment of Neo-Darwinian evolutionary theory as non-scientific led him to emphasize the theory as 'situational logic'. See Popper *et al.* (1974) 135. And it was his understanding of both Neo-Darwinism and falsificationism as situational logic (this term will be explained below) that first led him to identify the two. Later he came to treat Neo-Darwinian evolution as an empirical scientific theory. See Popper (1987) 139–55.
61. The following account is indebted to Sober (1993) Chapter One and also Purves and Orians (1984) Chapter Nine.
62. Sober (1993) 9.
63. Purves and Orians (1984) 934.
64. Ibid. 931.
65. Popper (1959a) 52.
66. Popper *et al.* (1974) 82.
67. Purves and Orians (1984) 933.
68. Ibid. 934.
69. Ibid.
70. Popper (1979) 146.
71. Popper (1990) 47.
72. Popper (1979) 146.
73. Ibid.
74. Popper (1990) 46–7. In this text Popper reveals the influence of Konrad Lorenz. See Lorenz (1977).
75. Popper *et al.* (1974) 133.
76. Popper (1979) 165.

77. Ibid. 31.
78. Ibid. 261.
79. Ibid. 266–9.
80. Ibid. 267.
81. Ibid. 242.
82. Ibid. 217.
83. Popper's account of language functions is indebted to Karl Buhler. See Buhler (1930).
84. In fact, all the above functions are also adaptations given Popper's solution to the problem of orthogenesis; see Popper *et al.* (1974) 138–41.
85. Although we will not examine it in any detail in this section, paralleling the relationship between higher and lower language adaptations is the relationship between the development of biological organs and exosomatic organs such as computers. More will be said about this in the next section in relation to the topic of objectivity.
86. Popper (1979) 239.
87. Popper *et al.* (1974) 1061.
88. Popper (1979) 106.
89. Ibid. 122–3.
90. Campbell (1974) 448–9.
91. Popper (1979) 25.
92. Popper *et al.* (1974) 449.
93. For more on Popper's implicit development of hypothetical realism, see Munz (1985).
94. Popper (1979) 292–3.
95. Popper *et al.* (1974) 149.
96. Popper (1959a) 30–1.
97. Popper (1979) 109. See Frege (1892). For more on the relationship of Popper to Frege, see Notturno (1985).
98. Popper (1979) 110.
99. Ibid. 108.
100. Ibid.
101. Ibid. 157.
102. Popper (1990) 17.
103. Ibid.
104. Popper *et al.* (1974) 1066–7.
105. Gilroy (1985).
106. Popper (1983) 48 and (1979) 25.
107. Popper (1979) 119–22.
108. Popper and Eccles (1977) 40.
109. Popper (1979) 109.
110. Ibid. 115.
111. Ibid.
112. Ibid. 116.

113. Popper *et al.* (1974) 138.
114. Ibid. 143.
115. Popper (1979) 114.
116. Ibid. 116–17.
117. Popper (1982a) 116.
118. Quine (1951).

Chapter 4

An Objective Social Order: Politics and Ethics

Introduction

In what follows, I treat Popper's ethical and political writings as aimed at exploring the conditions required for an objective social order. Throughout Popper's major writings on social and political philosophy, *The Poverty of Historicism* and *The Open Society and Its Enemies*, the problem of an objective social order is directly related to the problem of change. Fear of change and the desire to arrest it are the motivating factors behind the development of the historicist philosophies and totalitarianism Popper opposes, yet change in the form of the growth of knowledge is central to his response to historicism – the idea that historical laws of destiny determine socio-political developments.

'Objectivity' connotes the idea of distance, and in Chapter 1 I structured the political problem of objectivity in reference to the need for distancing political aims, practices, and procedures in a way that would make criticism possible and therefore secure objectivity. Thus, like his scientific methodology, Popper's theory of society and politics places intersubjective criticism at the center of his understanding of objectivity on the socio-political level. Moreover, the notion of distance plays a central role in Popper's characterization of an open society. For example, he argues it is only when humanity puts distance between normative injunctions and natural laws and accepts the former as the result of their own decisions that an open society that recognizes its own fallibility and freedom is possible. Nevertheless, Popper is not proffering the open society as a new utopian vision because problems are forever with us given human fallibility and because we lack any external guides to determine which goals and means of achieving them are the correct ones. For Popper, this means that we have to rely on ourselves and employ 'situational logic' to assess the changing demands of life in society.

Before sketching my approach in this chapter, the notion of distance employed in the formulation of the political problem needs addressing. 'Distance' as used in the formulation of the political problem has both a theoretical or abstract and a practical dimension. On the theoretical level talk about distance, for example between norms and facts, references a logical relation of deducibility and such abstract considerations are central to Popper's analysis. But Popper is also concerned with what might be called practical problems of distance; for example he argues one of the shortcomings of utopianism is that it advances policies that are so extensive in scope that they can't be subject to revision by the utopian engineers who initiate them. Although concerns over both senses of distance are present in Popper political thought, in what follows abstract theoretical dimensions are primary.

Proper appreciation of Popper's solution to the political problem of objectivity requires three aspects of his political thought to be addressed: (1) the background conditions for society, (2) the methodology of the social sciences, and (3) his critique of historicism. An erroneous take on the background conditions for society, on the methodology of social theory, and most especially on the value of historicism invites the merger between aims, procedures, and practices that undermines an open society. Finally, Popper never develops an ethical theory; however, in stark contrast to his own views on the matter I explore in the final section of the book how a genuinely Popperian ethics can be realized.

1. Evolution and the Myth of the Framework

I identify four theses in Popper's writings concerning the origin and provisions for society that underwrite his political theory. These are:

1. Human society has a pre-human origin, and our pre-human ancestors were social animals; therefore, society's origin is not to be found in a unique fact about human nature or psychology.[1]
2. Humans didn't make language, language made humans.
3. Agreement in intellectual frameworks is not essential to rational inquiry, and so not essential to political dialogue aimed at solving social problems. Cultural relativism is a non-issue for society and shouldn't prohibit social discourse.
4. A tradition of critical inquiry is possible despite the fact that shared intellectual frameworks are not essential to criticism.

The first thesis is advanced as part of Popper's critique of Mill's psychologistic sociology. Popper argues that Mill's psychologism is opposed to the genuine autonomy of sociology and forces Mill 'to adopt historicist methods'.[2] As noted in Chapter 1 above, Popper endorses Marx's epigram that asserts social existence determines human consciousness.[3] Popper's argues against Mill's psychologism and for the autonomy of sociology in a manner that emphasizes human society's evolutionary origins. Because psychologism reduces social rules and institutions to psychological facts about human nature and takes such facts as prior to social organization, it is forced

> [w]hether it likes it or not, to operate with the idea of a *beginning of society*, and with the idea of a human nature and a human psychology as they existed prior to society. ... It is a desperate position because this theory of a pre-social human nature which explains the foundation of society – a psychologistic version of the 'social contract' – is not only an historical myth, but also, as it were, a methodological myth. It can hardly be seriously discussed, for we have every reason to believe that man or rather his ancestor was social prior to being human.[4]

By treating society as a direct result of human psychology in the sense of human wants, needs and motivations there is an obvious tendency to conflate the social environment with the conscious actions of individual human agents.[5] This leads psychologistic sociologists to ignore the fundamental problem of sociology, 'the task of analysing the unintended social repercussions of intentional human actions'.[6] This point is addressed in the forthcoming section devoted to the methodology of social science; however, the more important point for the present discussion is that as an interpretive framework psychologism leads its practitioners away from seeing any distance between individual psychologically motivating factors and societal organization and so impedes forthright criticism of societal developments. If from the perspective of psychologism society is fundamentally a reflection of what individual agents know about themselves and intend in light of this knowledge, then the likelihood of societal practices, procedures, and aims standing in criticism of one another is diminished since they are all equally an expression of the agent's intentions. There is little need for a fallibilist mindset when a person can be comfortable with the fact that she is intimately aware of her own intentions and thus societal aims, procedures, and practices are for the most part mutually supportive because they express her design.

Evolutionary theory, however, incorporates as part of its theoretical structure the idea of random mutations that the organism itself is unaware of. Accordingly, our evolutionary heritage teaches us that there is not necessarily a ready-made fit between the environment and particular adaptations understood as incorporating 'aims'. And the lack of fit between an adaptation and the environment is equally true on a general level concerning an organism's survival. Therefore, adopting an evolutionary perspective helps the social theorist acknowledge that aims, practices, and procedures shouldn't be expected to be mutually supportive. Moreover, criticism by the social environment isn't to be feared because of the changes it may introduce; rather, problematic social contexts and the changes they introduce can be responsibly handled in light of the resources society provides us with. Many of these resources are the exosomatic tools of the descriptive and argumentative functions of language understood as evolutionary adaptations. In particular Popper identifies these tools with myths, artistic works, scientific theories, and the content of critical discussions (the denizens of World Three). The descriptive and argumentative functions of language through their use of the regulative notions of truth and validity allow for a type of feedback on conjectured solutions to problems and provide a biological analogue to the situational logic that Popper applies to social theory.[7]

Popper asserts that from an evolutionary perspective language is not of our making but made us, and this fact is of great importance to his social theory. Popper's account of the four stages of language (expressive, signaling, descriptive, argumentative) emphasizes the latter two as distinctively human. In particular, the argumentative function understood as an adaptation and therefore oriented to reconciling an organism with its environment is characterized by rational criticism. The object of this criticism is the objective content of the World Three denizens identified above. In the same way that any non-human organism grapples with its environment, so too humans struggle with the content of an intellectual world that is their own exosomatic product. The descriptive and especially the argumentative functions of language evidence how humans evolved to transcend or distance themselves from their merely biological origins. Fundamentally, Popper thinks this distancing can best be achieved by critically engaging the objective content of theories in a fallibilist way; however, this involves the recognition that unlike other organisms humans are not one with their theories. (Think again of the difference between Einstein and the amoeba.) Social science and practical politics can benefit from adopting the same fallibilist mindset when identifying theoretical aims, the arguments for them, and addressing issues of practice as well.

Popper recognizes it is a commonplace conclusion that a shared intellectual framework is requisite if social harmony and the rational discussion essential for societal development is to obtain. Popper attacks the idea of a shared framework as essential to social harmony and the quest for truth because he understands it to be a mainstay of relativism (conceptual and cultural), something he rightly identifies as a great threat to society and its development via the critical exchange of ideas.[8] In what follows, I argue that Popper and Donald Davidson, working from different perspectives concerning both the function of intellectual frameworks or what Davidson calls conceptual schemes and the nature of language, advance fundamentally the same position establishing the downfall of relativism.[9] Additionally, Popper's contribution enables us to appreciate better when it is acceptable to talk of divergent conceptual schemes, and how best to do so.

As noted in Chapter 1 above, the myth of the framework asserts 'a rational and fruitful discussion is impossible unless the participants share a common framework of basic assumptions or, at least, unless they have agreed on such a framework for the purpose of discussion'.[10] Popper's critique focuses on intellectual frameworks and 'not preconditions for discussions ... such as a willingness to share problems'.[11] Popper, unlike Davidson, thinks that frameworks/conceptual schemes are ultimately the object of rational inquiry. This is the case because he takes even observation statements to be theory-laden, thus conceptual schemes are present as interpretive structures at the most basic level of information gathering. The demand for shared intellectual frameworks underlies relativism in the sense that when this demand is both taken seriously and unrealized, people conclude that an objective and true account of reality isn't possible. This defeatist conclusion is supported by a host of corollary points that help account for why people are so easily persuaded to accept the myth and, consequently, relativism. First, the myth of the framework leads people to take agreement to be the aim of discussion. Because agreement can be secured in a non-rational fashion, for example, by force, no one can gainsay such a method since agreement as an aim says nothing about how minds meet but only demands that they do so.[12] And although the immediate rejoinder to this point is to claim that enlightened people obviously seek *rational* agreement, the demerit of this response is that in the hands of advocates of the myth 'rational' is equated with justificationism. This justificationist presupposition is expressed in 'the mistaken view that all rational discussion must start from some *principles*, or as they are often called, *axioms*, which in their turn must be accepted dogmatically if we wish to avoid an infinite regress'.[13] This point is satisfactorily addressed by Popper's critique of justificationism (see Chapter 1 above); however, because justificationism is central to the myth his anti-justificationism is not a purely academic matter but is immediately relevant to social life. Additionally, in the same justificationist vein the myth is supported by the demand for total agreement. In this case the idea is that unless the parties engaged in discussion come to agree on every point, understanding is impossible.[14] In the background of the demand for total agreement is an idealized form of justificationism, and, although Popper doesn't indicate it, a holistic account of truth as well. Third, people confuse fallibilism with relativism.[15] However, Popper argues that the two are actually unrelated because genuine fallibilism invokes truth in an absolute sense and not in a relativist manner, genuine fallibilism stresses 'that our theories can be absolutely false – that they can fall short of the truth'.[16] Moreover, the fact that at any time a person may be in error concerning her knowledge claims does not show truth to be relative since how a person knows X to be true and that X is true are completely separate issues. There is much else in Popper's paper that is worthy of consideration but I next analyze the grounds for the myth's failure and the mainstay of his critique of relativism.

The myth of the framework cannot escape a purely trivial formulation, even though people think the myth 'is a logical principle, or based on a logical principle',[17] and therefore is worth advocating. The myth states that a high quantity of shared content between participants in a discussion implies a high likelihood or probability for agreement among the participants. But since shared content in this context obviously references content that the participants mutually agree on, then all the myth states is that people are highly likely to agree to what they agree to. Because this is a vacuous principle, failure to satisfy it can't underwrite any philosophical position, even relativism. But even if we ignore the tautologous character of the above formulation it is still inadequate on Popperian grounds. Popper asserts an inverse relationship between probability and informative content; this means that as the content of a theory increases, its probability decreases. The end result is that the more people agree the less they say about the world. One final salvage attempt by the advocates of the myth seems possible. As it functions in the myth, the expression 'shared content' can be taken to denote both a broad and a narrow domain. The narrow domain references the set of shared discussion points all the discussion participants agree to, while the broad domain references the overall cognitive content the participants have at their disposal. Here the first domain is a subset of the latter. Applying Popper's insight about the inverse relationship between probability and content and keeping in mind the two domains of shared content, it follows that as the probability of agreement increases, content in one or the other or both domains decreases. Among these outcomes two are of special concern to advocates of the myth. First, if the domain of shared content decreases this would entail that the discussion participants have less content to work with, definitely disappointing but on the whole neither exceedingly damaging nor especially informative in a way that affects how advocates of the myth perceive it; however, a decrease in overall shared content is significant if its decrease occurs because content in the broad domain is transformed into content in the narrow domain, that is, portions of shared overall content are transformed into shared-agreed-to content. This scenario seems to be exactly what advocates of the myth of the framework desire, but in reality it is eminently undesirable. Reduction of overall content to shared-agreed-to-content takes us back to the first tautological formulation. Thus the end of the investigation is that under any formulation the myth is either vacuous, or falls prey to the inverse relationship between probability and content.

In contrast to the myth of the framework Popper advocates culture clash as key to fruitful discussion. Although Popper does not make the point in any direct way, his advocacy of culture clash is related to his insight that there exists an inverse relationship between probability and content. Fruitful discussions are the result of culture clash because the participants can gain novel insights from each other, i.e., an increase in informative content, the fewer points they agree upon between them. Simply put, if there are no differences between the participants in a discussion, what do they have to learn from one another? Popper states:

> I think that we may say of a discussion that it was the more fruitful the more its participants were able to learn from it. And this means: the more interesting questions and difficult questions they were asked, the more new answers they were induced to think of, the more they were shaken in their opinions... the more their intellectual horizons were extended. Fruitfulness in this sense will almost always depend on the original gap between the opinions of the participants in the discussion.[18]

Popper even goes so far as to state, 'had there been no Tower of Babel, we should invent it'.[19]

Popper's critique of relativism may receive greater appreciation by a wider audience if some parallels and some points of contrast to Donald Davidson's work on conceptual schemes are explored.[20] Davidson's work offers a serious critique of relativism grounded in the purported untranslatability of language as it is developed by the likes of Quine, Kuhn, and Feyerabend. These thinkers conclude the untranslatability of one language into another is evidence of divergent conceptual schemes, hence conceptual relativism. In response, Davidson acknowledges that it is not self-evident that to be a language is to be translatable; however, he thinks this fact can emerge as a conclusion to an argument that examines the possibility for total or partial failures of translation.[21]

Davidson asserts that the failure of translation thesis doesn't adequately take into account an important fact about language: to identify someone as a language user is very similar to attributing to an individual a set of attitudes 'such as belief, desire, and intention',[22] and this attribution is grounded in a unique relation and 'until we can say more about what this relation is, the case against untranslatable languages remains obscure'.[23] Davidson's transition from this point to his next set of topics is not clearly spelled out, but his general point seems to be something like the following. Attributing a belief to someone seems to involve ascribing to that person a conceptual scheme and a relation to an object of belief. This dualism between scheme and content invites two possible interpretations indebted respectively to Strawson and Kuhn.[24] The Strawson-based perspective considers a relation between scheme and content where the framework of concepts is fixed and we apply them to a plurality of distinct objects of inquiry, say, various possible worlds. The Kuhnian perspective treats the content as fixed, but it is open to appraisal in terms of diverse conceptual schemes. It is the Kuhnian perspective that is the principal object of inquiry as a source of conceptual relativism. Concerning it Davidson argues that the dualism between scheme and content is erroneous. First, there is no 'uninterpreted reality' so the idea of a neutral content that invites various interpretive schemes and therefore conceptual relativism is incorrect;[25] moreover, Davidson argues that to talk of another person as having a language is to do so in terms of one's own language. In ascribing language to another person one has no choice but to take one's own language and framework for belief attribution as a standard. Davidson's approach is indebted to Kant. In the same way that Kant's transcendental deduction aims to establish an objective world beginning from and assuming nothing more than the knower's perspective on how things appear to her, so too Davidson argues that an objective account of a speaker's language, indeed any language, takes its start from the interpreter's perspective and provides the only constraint and the only insight into what constitutes a language that a person can avail herself of. Davidson identifies this perspective with charity, arguing that translation is only possible if the interpreter assumes that the speaker whose language is being interpreted most often gets things right.[26] Having both relinquished the dualism between scheme and content and established the conditions for the possibility of translation, Davidson concludes:

Without the dogma, this kind of relativity [conceptual] goes by the board ... In giving up the dualism of scheme and the world, we do not give up the world, but re-establish unmediated touch with the familiar objects whose antics make our sentences and opinions true or false.[27]

Popper would applaud Davidson's attack on relativism and agree with him concerning the theory-ladenness of statements, but I think Popper has a better understanding of how conceptual schemes function in our intellectual endeavors. Consistent with his conjectural or hypothetical realism, Popper can advance conceptual schemes as essential to a critical methodology. Inquiry proceeds by advancing conceptual schemes of high testability, and inquirers must be open to the prospects that another theory of greater testability and explanatory power can replace initial guesses. Popper likened frameworks to prisons and argues that intellectual life is a fight for freedom but only into larger and larger prisons.[28] Perhaps this metaphor is unfortunate because it invites the idea of conceptual relativism by allowing for a plurality of frameworks to choose between. However, I think in the background of Popper's analogy is the aim of making sense out of tradition, and in particular a tradition of critical inquiry. Here Davidson can help. Conceptual relativists are content to treat historical epochs as isolated and to explain the transition from one to another as revolutions occasioned by gestalt-like switches in worldview. In this way they basically walk away from any responsibility to explain how the transition from one epoch to another takes place in a rational manner. As it concerns the domain of intellectual inquiry Popper's answer is that the transition occurs via a tradition of critical inquiry. His account of tradition can be supplemented by Davidson's analysis of conceptual schemes that denies any terminal cutoff point to distinguish frameworks in a way that denies any continuity between them; in this way the idea of civilization as critical dialogue makes good sense.

2. Social Science Methodology

Popper argues for both a unity of method concerning the natural and social sciences and for some important differences.[29] Both the natural and the social sciences offer 'deductive causal explanations',[30] structure their conjectures as falsifiable, and begin with problem situations. As well, theories in both sciences produce testable consequences and thus allow for corroborated theory preferences. However, the social sciences differ most dramatically from the natural sciences because 'in most social situations there is an element of *rationality*'[31] and this element of rationality invites the use of theoretical models that enable social scientists to simplify observed social situations. Models, according to Popper, are present in both the natural and the social sciences, but they are more appropriate to the social sciences because in most cases the social sciences lack explanatory universal laws that in conjunction with initial conditions can generate singular predictions. Lacking universal laws, social science inquiry is more about an investigation into initial conditions and thus situational models best provide a framework for inquiry.[32]

The situational models social scientists construct rely upon a zero method.[33] The zero method is a

> method of constructing a model on the assumption of complete rationality (and perhaps also on the assumption of the possession of complete information) on the part of all the individuals concerned, and of estimating the deviation of the actual behaviour of people from the model behaviour, using the latter as a kind of zero co-ordinate.[34]

The great value of the method of situational analysis or situational logic is that it enables social scientists to address the fundamental problem of social theory: how to analyze the unintended consequences of intentional human action.[35] Compare this approach to historicist methods. The historicist appeal to historical laws contradicts the notion that there are any unintended consequences characteristic of the social environment. In this way, genuine novelty is not possible, and no facts about the social world escape the leaders who grasp the laws of historical destiny. What is not at issue here is the prediction of 'the precise results of a *concrete* situation, such as a thunderstorm, or a fire', since Popper recognizes that this is beyond the capability of even the natural sciences.[36] However, the historicist approach to sociology must allow neither unintended consequences nor for them to appear to be unforeseen in a way that would necessitate change in the societal blueprint, since this would be identical with an admission of a loss of control on the part of the power elite. However, the method of situational analysis by means of the zero method and by taking as its object changing initial conditions does not introduce substantive elements grounded in either metaphysics (Plato, Hegel) or scientism (Marx) or psychologism (Mill), but at an abstract level tries to capture 'the objective social situation'.[37] Thus the objective social situation is a consequence of realizing that sociological laws are not to be had and social institutions and psychological aims and motivations are to be transformed into objective facts of the situation. In this way, Popper believes a social scientist can make sense of placing herself in the shoes of Charlemagne and conclude; 'admittedly, I have different aims and I hold different theories ... but had I been placed in his situation thus analysed – where the situation includes goals and knowledge – then I, and presumably you too, would have done what he did'.[38]

This approach is formal and abstract and thereby has the advantage of allowing social life as an object of inquiry to speak for itself and to be analyzed on its own terms with no substantive assumptions; instead it requires only a minimum formal requirement expressing a thin notion of rationality that asserts rational agents act 'adequately to the situation'.[39]

Applying Popper's methodology of situational logic to the problem of objectivity in politics we can see it invites aims, practices, and procedures to be treated as distinct since the unforeseen consequences of human actions may be the result of, say, aims conflicting with procedures. Thus to demand that the three be one is to miss learning from the objective social situation in the name of securing metaphysical, scientific, or psychologistic frameworks that are comfortable or familiar and provide easy answers for societal living.

3. Historicism

Popper's critique of historicism and the totalitarian political theorists that make use of historicist philosophies is one of the most important moments in the history of political theory. However, its merits shouldn't be judged solely as critique; rather Popper's intent, especially in *The Open Society and Its Enemies,* is to bolster an alternative – democracy. But this entails that democracy and totalitarianism are fundamentally different. This difference Popper explains by asserting that democracies can remove their leaders without bloodshed, while political change in tyrannical states requires armed violence.[40] Nevertheless, there is more to the difference than just this distinguishing element because the violent as opposed to non-violent alternative reflects a fundamental difference at the level of methodology. This difference is located in how political aims, practices, and procedures are informed by fallibilism and the role critical inquiry plays as a response to human fallibility.

Totalitarian leaders invoke historicism and exploit its meager intellectual resources to secure authority as prophets of the future and as agents of social policy. To admit to error is to invite criticism and so undermine authority.[41] The future-directed nature of historicism gives it the means to forestall criticism by explaining away incongruities and possible refutations of the societal framework because all errors will be reconciled in the future.[42] This clears the path for leaders who claim to know the historical laws of destiny to steer society towards its ultimate end by means of utopian engineering. The methodology of utopian engineering requires the unity of aims, practices, and procedures as supportive of tribalism and thus a closed society.[43]

In fact, those people who assert the inevitability of totalitarianism argue that democracy in its fight against totalitarianism is forced to adopt totalitarian practices, thus whether intended or not, end and practice merge.[44]

The central methodological difference between democracy and totalitarianism is found in how the former incorporates fallibility and a corresponding critical mindset into the relationship between aims, practices, and procedures. Totalitarianism allows for the criticism of practices and procedures but only insofar as it is directed toward improvement of the aim; in this way the aim can never undergo scrutiny in a way that allows it to be perceived as erroneous. Democracy recognizes the need for criticism at all levels of societal organization and criticism is most effective when the consequences of our plans are near at hand and don't outrun our ability to control them. Therefore, Popper argues for the merits of piecemeal social engineering in opposition to utopian planning. The latter attempts to avoid unforeseen consequences by appealing to historicist principles and by destroying anything that fails to conform to the social blueprint made in conformity with historical laws of destiny. Piecemeal social engineering advances tentative agendas that are open to criticism by allowing aims, practices, and procedures to stand in judgment on one another so to determine the merits of a particular policy. Popper recognizes that his theory of piecemeal social engineering contrasts in an important way to his general theory of method:

> And it is a fact that my *social theory* (which favours gradual and piecemeal reform, reform controlled by a critical comparison between expected and achieved results) contrasts with my *theory of method*, which happens to be a theory of scientific and intellectual revolution.[45]

What is at issue here can best be understood by examining a point of fundamental difference between Kuhn and Popper. Kuhn's account of science emphasizes a difference between revolutionary and normal periods of science.[46] Normal periods of science are characterized by a set framework or paradigm for problem solving. When anomalies arise within the paradigm, accepted resources for problem formulation and problem solving are unavailable and this sets the stage for a revolutionary change in worldview. Concerning the natural sciences, Popper understands everyday science to be revolutionary science; in a sense there is no normal science because the scientist is advancing unique solutions to problems and encountering unanticipated refutations of his proposals. Granted, these proposals and their refutations might not be groundbreaking enough to change the course of a discipline, but they are novel nevertheless. However, on the political scene Popper thinks that our proposals should avoid revolutionary attempts to change the fabric of society, because, as noted above, revolutionary proposals in the social sphere can escape our ability critically to control them.

Popper's emphasis on critical control of revolutionary agendas leads him to acknowledge the paradox of sovereignty.[47] Can our political proposals and our drive to be critical lead us radically to reject democracy and criticism? Popper's response is twofold. First, to ask this question is to take seriously the question 'Who should rule?' and ascribe an authoritative status to the decision of the ruling party (king, majority of voters, etc.) such that if they say democracy and a critical mindset should be abolished it would have to be so. However, Popper suggests that the question of who should rule must be replaced by the question 'How can incompetent rulers be gotten rid of?' Nevertheless, it seems that our sovereignty permits us to reject this question as well. To my knowledge Popper never directly considers this scenario. However, rejecting this question is analogous to an individual refusing to jettison ineffective decision-making processes on the personal level. Only neuroses or a commitment to deny one's own fallibility (perhaps itself a neurosis) could motivate such a decision. Thus, only by denying our own fallibility can such an idea get off the ground. Unfortunately, historicism has supported a belief in just that.

4. Ethics

My aim in this section is not to identify amidst Popper's sometimes contradictory and farraginous writings and comments on ethics what could pass for a Popperian ethical theory by comparing it to established moral theories and thereby situate his views in reference to the thought of others; rather, my goal is to sketch a genuine Popperian ethics employing the full resources of his treatment of objectivity. This does not mean that his writings on ethics are entirely irrelevant to this undertaking, and it does not mean that the views of other philosophers cannot be of assistance to clarify what is at issue; nevertheless, neither is an adequate guide to what is best in his thought – something new is afoot. However, before exploring what I take to be the revolutionary implications of Popper's thought for ethics, two points need to be addressed.

First, Shearmur has carried out the project of situating Popper's ethical views in relation to the thought of other ethical writers in an exciting and important way; but space considerations do not allow me to treat all of his valuable insights. However, three central elements identified by Shearmur must be recognized: (1) Popper's epistemological insights extend to his views on ethics; (2) there are important parallels with Kantian ethics; and

(3) comprehensively critical rationalism (CCR) can make an important contribution to ethical theory.[48]

Second, to label Popper's writings and comments on ethics mixed is accurate; to charge him with contradictions appears tendentious, perhaps irresponsible, if the charge cannot be made to stick. Briefly, consider that Popper asserts the autonomy of ethics, and he means not only that no fact can entail any value, but, as well, ethical theory is unable to inform moral decisions in any way because it is 'hot air'.[49] However, he then goes on to advance his own ethical views reflective of negative utilitarianism. This seems like both an appeal to facts – reduce suffering – and to invoke an ethical theory. Popper's plea for the removal of unnecessary suffering need not be treated as incorporating an ethical theory, since it may be possible he is advancing an ethical view without an appeal to principles.[50] Still, he asserts ethical values to be inhabitants of World Three, and because he rejects emotivism he would reject the notion that values are mere descriptions, thus values would at least have to be abstract in a way that parallels theories. The end result is that Popper both dismisses ethical theorizing and advances what looks like an ethical theory that is to be treated as a denizen of World Three.

As stated in Chapter 1 above, I take the fundamental question of ethics to be 'Why be moral?' This question in turn has led philosophers to reflect on the relationship between reason and morality; thus philosophers ask, 'Is it rational to be moral?' I aim to sketch how a Popperian philosophy can best answer these questions. Central to the response is the revolutionary theory of reason developed by Popper and his followers (Bartley, Miller, *et al.*): CCR. This theory provides a framework for a 'thin' notion of rationality able both to explain the relationship between reason and morality and to provide a resource for generating particular normative rules. In the discussion that follows, I understand by morality a personally chosen system of constraints on the pursuit of individual self-interest,[51] and although this notion is less rich than, for example, teleological accounts of morality or accounts structured in terms of the good, I hope to show that in conjunction with the thin notion of rationality to be developed here, this restricted take on morality is sufficient to account for our commonplace moral commitments.

Ethical theory in the Western philosophical tradition is fundamentally flawed because it works with a faulty account of reason. Attempts to explain the relationship between reason and morality have taken a justificationist form, but because there is no such thing as justification in either the theoretical or practical realm, the result is a long-standing failure to make sense out of our ethical lives. For the critical rationalist, the reasonable person is the person who advances criticizable conjectures in an attempt to solve a particular problem. Rational enquiry proceeds by attacking the conclusion of an argument to show it to be false, and if the conclusion is false this falsity is logically retransmitted to at least one of the premises from which it is derived. Additionally, because the premises of an argument in no way support or justify its conclusion, it is no great achievement to attack them because they could never meet the justificationist demands placed on them anyway (it may be psychologically satisfying to attack the premises of an argument but the serious inquirer into truth will not waste her time on irrelevancies), thus the target of criticism is the conclusions people advance.[52] This fact saves us the expenditure of having to invest in an initial attack on the premises of an argument, and, as argued earlier in the text, Popper's negative methodology enables inquirers to identify a critical preference between theories.

Rather than ask what reason is, a Popperian ethics adopts Popper's practice of replacing a substantive approach to philosophical issues with a methodological one, as he did in the cases of epistemology and causality. This means that, wherever possible, the problems of ethics should be replaced by methodological issues. A Popperian ethics shows rational methodology and moral methodology to be coextensive; therefore to be moral is to be rational.

The demand for criticizability as characteristic of rationality is commensurate with the demand for fairness in the ethical domain. No person can fairly ask another individual or group to perform an action or adopt a statement as true without them having, or in some cases providing them with, the resources to be critical of the request at issue. At this point in most discussions of ethics, the norm is: (1) to ask what makes criticizability or the recourse to criticism characteristic of morality, and (2) to identify instances where it seems fair systematically to deny oneself or others the resources to be critical of the statements or policies that are advanced for acceptance. The latter search for counterexamples is praiseworthy, but the former request is not. From a Popperian perspective, to the extent (1) invites the open-question argument and thus the naturalistic fallacy, it begs an essentialist answer and so, like all essentialist postures, it forestalls inquiry. Moreover, because it implies that a basis for calling something moral must be provided to distinguish it from, say, a command that is prudential in an instrumental sense, it invites foundationism. Finally, it fails to appreciate the conjectural quality of all our theories including ethical ones. The thesis advanced here concerning the relationship between criticism and morality is a conjecture and, like any conjecture, it does not have to prove its pedigree to show itself worthy for rational discussion, because this would be justificationism.

Consequently, the question contained in point number (1) above encourages essentialism, foundationism, and justificationism, each a philosophical position Popper rejects as part of any true account of both objecivity and rationality. However, to dismiss foundationism in the above manner seems to deny to ethics the task of demarcating itself from other subject matters, and thus to deny to ethical theory an undertaking central to Popper's own philosophical project. But even though I believe the charge of foundationism is not wide of the mark, the Popperian ethics developed here can nevertheless satisfy the traditional demand to account for the distinctiveness of ethics. To begin, consider the following passage from Miller:

> There is no great harm done if we expand the domain of criticizability so that, quite trivially, all statements are criticizable. But CCR wants more: the truth of a statement has to be something that can be critically investigated. Without that there is no comprehensively critical rationalism. Thus I reject the thesis that criticizability is an automatic property of all statements. It is not an intrinsic property of statements at all, but an honour that must be bestowed on them by the development of appropriate methods of criticism.[53]

As I see it, Miller's point is that the criticizability of a statement can't ignore the statement's content, but its content is not all there is to a statement's criticizability. Criticizability is also a function of what a person does, whether she adopts immunizing strategies towards her theories, etc. Thus to capture this interplay between the properties of a statement and the disposition one takes towards it, I introduce the term 'positions' as specifying a domain reflective of both elements. Now more often than not, ethicists have attempted to account for the distinctiveness of ethics by establishing a unique aim, for instance the complete life of virtue, or the greatest good for the greatest number; or they have appealed to an exclusive subject matter, for example, the logical form of moral statements or the workings of moral epistemology. Nonetheless, these approaches are wrong. The distinctiveness of ethics is to be found in the methodological role that criticism plays in the formulation of a moral position. This can be seen in the following two ways. First, criticism directed at any subject matter other than ethics aims to exhaust the informative content of the subject matter at issue; thus, there is a law of diminishing returns concerning factual and prudential matters because in the name of rigorous testing criticism eventually leads to tests of lesser severity as unsullied criticizable content is depleted.[54] But there is no law of diminishing returns for ethics; exposing a position to criticism doesn't introduce conditions of decreased fairness; rather the criticizability of a position is commensurate with its fairness. Thus morality expresses an asymmetry that is not true of factual and prudential domains subject to critical inquiry.

Next, consider the following sets of contrapositives:

> If a position is false, then it is a criticizable position. ::
> If a position is uncriticizable, then it is a true position.
> If a position is fair, then it is a criticizable position. ::
> If a position is uncriticizable, then it is an unfair position.

Comparing the two sets, the first conditional statement in the first set is true, while its contrapositive is indeterminate. That a position is uncriticizable tells us nothing about its actual truth; although it does tell us something about our ability rationally to assess it. However, in the second set of conditionals criticizability is inseparable from the truth of the statements. It is not possible to hold a position fairly and to make the position in principle immune to criticism. Thus criticizability always informs what we take to be moral, but criticizability does not fix the truth or falsity of statements. Thus those areas of inquiry or domains that consider truth or falsity alone are distinct from moral positions in reference to how they relate to criticizability.

In light of the emphasis a Popperian ethics gives to criticism, a central question that arises is how are moral statements subject to reasonable constraints or refutation? Because values arise in the context of solving problems it seems obvious that the facts or states of affairs that gave rise to the problem situation should inform the choice between values that confronts us.[55] Nonetheless, philosophers have badly construed the relationship between facts and values. Critical rationalists, because of their explorations into the nature and limits of deductive logic, are best situated to provide some illumination. Popper accepts the dualism of facts and values; no fact can validly entail a value.[56] However, Bartley's analysis of the transmissibility assumption characteristic of deductive inferences shows that there is an argument that is even more basic to much of traditional ethical theory and its outcome perhaps more devastating to the relationship between facts and values.[57] Bartley argues that not every property of the premises of an argument is deductively transmitted to the conclusion. He reasons that because any statement validly implies a tautology, and because tautologies are devoid of all content, the empirical content (Bartley calls it the 'empirical character') of the premises of an argument is not deductively transmissible to its conclusion even when the premises are true. Although not emphasized by Bartley, this point can be generalized to include all descriptive content, and this fact has previously unrecognized implications for ethical theory.

Hume found the deduction from facts to values indefensible because the conclusion of such an inference contains a genuinely new relation not found in the premises, thus the inference is invalid.[58] But Bartley's analysis of the transmissibility assumption reveals that even when the inference is *valid* the descriptive content of the premises is deductively unrelated to the conclusion, that is, it's not deductively transmissible, and so not necessarily related to any conclusion we may want to establish, moral, factual, or otherwise.

Hume:	Bartley:
Fact	Descriptive content
Fact	Descriptive content
Value	Tautology
Invalid: a new relation in the conclusion that is not in the premises	Valid: however the descriptive content is not deductively transmissible

This point, which is not unconnected to the general fact that the premises of an argument never serve to justify or establish its conclusion even when they entail it,[59] suffices to close the door once and for all on any ethical theory that aims to derive prescriptive conclusions logically from descriptive premises. Remember, the whole point of such an inference is to establish that the descriptive content of a statement is necessarily related to the prescription asserted to follow from it, and thereby the descriptive content is somehow logically relevant to the moral judgment a person accepts as obligatory. However, the counterexample serves to show that there is no deductive transmission of descriptive content from the premises to the conclusion. When this point is accepted along with the fact that there are no inductive inferences and no inductive probability the end result is that the relationship between descriptive content and prescriptive judgment is at best psychologically relevant, but thereby arbitrary. An extreme rejoinder asserting that no one expects the descriptive content of a statement to transmit to a moral statement because complete transmission of descriptive content would thereby transform the moral statement into a descriptive one only serves to highlight the problem of logically deriving prescriptive claims from descriptive ones. Indeed, if any such transmission of descriptive content seems problematic in the instance where the logical relationship is a necessary one, this fact only highlights the logical problem of why the deductive relationship between descriptive and prescriptive claims is desired in the first place.

If it is a good thing that the two are not necessarily related because the necessary relationship impugns the integrity and autonomy of moral statements, then the dualism of fact and values is a credit to moral inquiry. This, of course, is exactly Popper's point, and strikingly it obtains when the inference at issue is valid.

Does the above argument show that there is *no* deductive relationship between descriptive statements and moral injunctions? Consistent with Popper's general methodology of conjectures and refutations, Bartley argues that facts relevant to a moral context can constrain it in a negative way.[60] Concerning the maxim that ought implies can, Bartley asks under what conditions can we say to Jones that he ought to be a genius. He argues that if Jones is mentally handicapped and for physiological reasons unable to attain such an end, then the physiological facts serve to constrain the moral injunction that Jones ought to be a genius. Although the example is a little strained, since to say to Jones that he ought to be a genius permits the 'ought' to be construed as prudential, still the general point is that facts can negatively constrain value judgments. Nevertheless, ought statements are not the only class of moral judgments, and Bartley denies that the above account of how experience may regulate a moral position is exhaustive of all moral judgments. This invites the question whether there is a more general constraint on the moral positions we adopt. The Popperian answer is yes and the constraint is to be found in the demand for criticizability itself. Somewhat parallel to Kant's categorical imperative[61] and Rawls' notions of reflective equilibrium and justice as fairness,[62] a Popperian ethics asserts that criticizability is recursively present in the moral values that arise as constraints on the pursuit of self-interest as human agents attempt to solve the various problems that confront them. Morality requires that no person can ask herself or others to adopt as *moral* a position that is in principle immune to criticism. Morality requires fairness and fairness requires criticizability. Criticizability is recursively present because it initially occurs in any moral position that we formulate and is non-circularly incorporated in any extensional development of that position. That moral positions are structured in reference to criticizability is evident in our common moral commitments. Lying, murder, and theft each undermine the person who is lied to, murdered, and robbed from criticizing the wrongs committed against them. This is obviously the case in the first two instances. The liar aims to impair the ability of the person or group lied to from critically evaluating her statements, from getting to the truth. A person who is murdered obviously is denied any recourse to criticism concerning the evil committed against her. The case of theft is slightly different because a thief can taunt his victim with what is stolen from her and so not all avenues of criticism are closed; however, the thief cannot remain a thief and leave every avenue of refuting the theft open. Kantian philosophers obviously will recognize in this exposition parallels to Kant's contradiction-in-thought test that avails itself of a recursive testing procedure to determine whether a subjective maxim can be raised to the status of a law. Kant's methodology emphasizes universalizability and publicity. If lying is made into a law of nature so that everyone necessarily lies and this is publicly known, then lying is futile.[63] A Popperian ethics is asking us to recognize that moral positions are formulated in reference to criticizability and if a position is immunized from criticism, then the position is unfair, hence immoral, and opposed to reason. Here, ethics is not reduced to a principle; rather it invites a methodology that investigates both the content and logical form of the statements that make up the moral position advanced, and it also asks us to consider how the statements are held, what position an agent adopts towards them.

Thus a Popperian ethics is a far cry from a sterile proceduralism. It requires that we establish social institutions that are open to criticism and that promote criticism throughout the public sphere. To do so requires that we re-evaluate our understanding of privacy and the boundaries that go with our various social roles concerning what is accessible to the scrutiny of others. On a personal level a Popperian ethics challenges us to enter into the lives of others in a manner commensurate with the respect for persons. To welcome criticism from another and to advance criticisms of her position is to avoid authoritarian postures and to acknowledge her views as worthy of concern. This does not mean that there is no place for authority in the personal and public arena, but only that the demand for authorities to behave with fairness will be informed not by the idea that they justify their positions to us, but by the requirement that they have formulated their positions in a way that exposes them to elimination.

Finally, the ethical theory sketched above can account for our interest in morality. Criticism is of interest to us because we know ourselves to be fallible. Even those persons who would choose to ignore this fact must confront a world that is often recalcitrant to even the best-made schemes. Thus it is the objective world in all its overwhelming force that commands the most ardent subjectivists to reject megalomania and see the value of criticism. Ultimately, to give the world its due, and to reckon it as a force to reckon with, is part of being fair to oneself; and like self-love which is required if we are to love others, fairness to oneself is the taproot for the fairness we extend to others as we constrain the pursuit of our self-interests in the attempt to solve the problems that arise as part of making our way in the world. In doing so, we are forced to acknowledge other persons as contributors to the problems we confront, as well as helpmates in the conflict. Socially situated, conscious of our fallibility, proceeding by means of controlled guesswork, we look to one another to build a social order where the idiosyncrasies of self-interest are subject to objective constraints. For the Popperian ethics sketched here, this means that we bring into play all the objective metaphysical, political, epistemological, and linguistic insights at our disposal; not just in the name of building a system of controls, but as a basis for a not undue humility.

Notes

1. Popper (1945b) 93.
2. Ibid. 92.
3. Ibid. 89.
4. Ibid. 93.
5. Ibid.
6. Ibid. 95
7. Popper (1979) Chapter 6.
8. Popper (1945b) 369ff.
9. Davidson (1984b)
10. Popper (1997) 34–5.
11. Ibid. 35.
12. Ibid.
13. Ibid. 59.
14. Ibid. 33.
15. Ibid. 48.
16. Ibid.
17. Ibid. 35.
18. Ibid. 35–6.
19. Popper (1963) 352.
20. Davidson (1984b).
21. Davidson (1984b) 185–6.
22. Ibid. 186.
23. Ibid.
24. Ibid. 187.
25. Ibid. 198
26. Ibid. 197.
27. Ibid. 198.
28. Popper (1997) 53, 61.
29. Popper (1957) 130ff.
30. Ibid. 131.
31. Ibid. 140.
32. Popper (1997) 165–6.
33. Popper (1957) 141ff.

34. Ibid. 141.
35. Popper (1945b) 95.
36. Popper (1957) 139.
37. Popper (1997) 167 and (1996) 79.
38. Popper (1996) 79.
39. Ibid. 177.
40. Popper (1945a) 4 and (2005) 94.
41. Popper (1945a) 189.
42. Ibid. 9.
43. Ibid. 172–3.
44. Ibid. 2.
45. Popper (1997) 68.
46. Kuhn (1970).
47. Popper (1945a) 123–4.
48. Shearmur (1996) Chapter 4.
49. Popper *et al.* (1974) 154.
50. Dancy (2004).
51. Gauthier (1988).
52. Miller (2006) 78–9.
53. Miller (1994) 86.
54. See Chapter 2 above.
55. Popper *et al* (1974) 154–5.
56. Popper (1945a) Chapter 5.
57. Bartley (1984) 261–5.
58. Hume (1978) 469.
59. Miller (2006) 68–76.
60. Bartley (1984) 199ff.
61. Kant (1956).
62. Rawls (1971).
63. Kant (1993) 15.

Bibliography

Agassi, Joseph (1993). *A Philosopher's Apprentice: In Karl Popper's Workshop*. Amsterdam: Rodopi.
—— (1995). 'The Theory and Practice of Critical Rationalism'. In *The Problem of Rationality in Science and Its Philosophy*, ed. Jósef Misiek, Dordrecht: Kluwer Academic Publishers.
Albert, Hans (1999). *Between Social Science, Religion and Politics*. Amsterdam: Rodopi.
Artigas, Mariano (1999). *The Ethical Nature of Karl Popper's Theory of Knowledge*. Berne: Peter Lang.
Ayer, Alfred Jules (1952). *Language, Truth and Logic*. New York: Dover Publications, Inc.
Barnes, Eric (1995). 'Approximate Causal Explanation'. *Philosophy of Science*, 215–26.
Bartley, William W. (1976, 1978, and 1982). 'Critical Study of the Philosophy of Karl Popper', *Philosophia*, Parts I–III.
—— (1983). 'Non-Justificationism: Popper versus Wittgenstein'. *Proceedings of the 7th International Wittgenstein Symposium*. Vienna: Holder–Pichler–Tempsky.
—— (1984). *The Retreat to Commitment* (revised 2nd edn). La Salle: Open Court.
—— (1987). 'Philosophy of Biology vs. Philosophy of Physics'. In *Evolutionary Epistemology, Theory of Rationality, and the Sociology of Knowledge*, ed. Gerard Radnitzky and William W. Bartley III. La Salle: Open Court.
Binns, Peter (1978). 'The Supposed Asymmetry between Falsification and Verification'. *Dialectica* 32: 29–40.
Blackwell, Richard J. (1991). *Galileo, Bellarmine and the Bible*. Notre Dame: University of Notre Dame Press.
Bloor, David (1976) *Knowledge and Social Imagery*. London: Routledge and Kegan Paul.
Boon, Louis (1979) 'Repeated Test and Repeated Testing: How to Corroborate Low-level Hypotheses'. *Zeitschrift für allgemeine Wissenschaftstheorie* 10: 1–10.
Buhler, Karl (1930). *The Mental Development of the Child*. London: Oxford University Press.
Campbell, Donald (1974). 'Evolutionary Epistemology'. In *The Library of Living Philosophers*. Vol. 14, *The Philosophy of Karl Popper*, ed. Paul A. Schilpp, 413–63. La Salle: Open Court.
Catton, Philip and Graham Macdonald (eds) (2004). *Karl Popper: Critical Appraisals*. London: Routledge.
Chesterton, G. K. (1957). *Orthodoxy*. London: The Bodley Head.
Corvi, Roberta (1997). *An Introduction to the Thought of Karl Popper*. London: Routledge.
Curie, Eve (1938). *Madame Curie: A Biography by Eve Curie*. Translated by Vincent Sheean. New York: Doubleday, Doran and Company, Inc.

Dancy, Jonathan (2004). *Ethics Without Principles*. Oxford: Clarendon Press.
Davidson, Donald (1984a). *Inquiries into Truth and Interpretation*. Oxford: Clarendon Press.
—— (1984b). 'On the Very Idea of a Conceptual Scheme'. In Davidson (1984a), 183–98.
Deluty, Evelyn W. (2005). 'Wittgenstein's Paradox: *Philosophical Investigations*, Paragraph 242'. *International Philosophical Quarterly* 45: 87–102.
Derrida, Jacques (1982) *Margins of Philosophy*. Translated with additional notes, by Alan Bass. Chicago: University of Chicago Press.
—— and John D. Caputo (1997). *Deconstruction in a Nutshell*. New York: Fordham University Press.
Einstein, Albert (1932). *Relativity: The Special and the General Theory. A Popular Exposition*. London: Methuen & Co.
Eiseley, Loren (1961). *Darwin's Century*. New York: Doubleday.
Fitch, G. W. (2004). *Saul Kripke*. Montreal: McGill-Queen's University Press.
Frege, Gottlob (1892/1966). 'Sinn und Bedeutung'. In *Zeitschrift für Philosophie und philosophische Kritik,* n.s. 100: 25–50. / 'On Sense and Reference'. In Peter T. Geach and Max Black (eds), *Translations from the Philosophical Writings of Gottlob Frege*. Oxford: Blackwell, 1966.
Gatens-Robinson, Eugenie (1993). 'Why Falsification is the Wrong Paradigm for Evolutionary Epistemology: An Analysis of Hull's Selection Theory'. *Philosophy of Science* 60: 535–57.
Gauthier, David (1988). *Morals by Agreement*. Oxford: Clarendon Press.
Gendler, Tamar Szabó and John Hawthorne (eds) (2002). *Conceivability and Possibility*. Oxford: Oxford University Press.
Gibson, Roger F. (ed.) (2004). *Quintessence: Basic Readings from the Philosophy of W. V. Quine*. Cambridge: The Belknap Press of Harvard University Press.
Gillies, Donald (1993). *Philosophy of Science in the Twentieth Century: Four Central Themes*. Oxford: Blackwell Publishers.
Gilroy, John D. Jr. (1985). 'A Critique of Popper's World 3 Theory'. *The Modern Schoolman* 62: 185–200.
Goodman, Nelson (1973). *Fact, Fiction, and Forecast* (3rd edn). Indianapolis: Bobbs-Merrill.
Gorton, William A. (2006). *Karl Popper and the Social Sciences*. Albany: State University of New York Press.
Hacohen, Malachi Haim (2002). *Karl Popper – The Formative Years, 1902–1945*. Cambridge: Cambridge University Press.
Hamrum, Charles (ed.) (1983). *Darwin's Legacy*. Cambridge: Harper & Row.
Hesse, Mary (1974). *The Structure of Scientific Inference*. London: Macmillan Press.
Hoyningen-Huene, Paul (1993). *Reconstructing Scientific Revolutions*. Chicago: University of Chicago Press.
Howson, Collin and Peter Urbach (1989). *Scientific Reasoning: The Bayesian Approach*. La Salle: Open Court.

Hume, David (1975). *An Enquiry Concerning Human Understanding*. In *Enquires Concerning Human Understanding and The Principles of Morals* (3rd edn), ed. L. A. Selby-Bigge, revised by P. H. Nidditch. Oxford: Clarendon Press.

—— (1978). *A Treatise of Human Nature*. L. A. Selby-Bigge edn, Oxford: Oxford University Press; P. H. Nidditch edn, Oxford: Clarendon Press.

Jarvie, Ian C. (2001). *The Republic of Science: The Emergence of Popper's Social View of Science 1935–1945*. Amsterdam: Rodopi.

Jarvie, Ian and Sandra Pralong (eds) (1999a). 'Popper's Ideal Types: Open and Closed, Abstract and Concrete Societies'. In Jarvie and Pralong (1999b), 71–82.

—— (1999b) *Popper's Open Society after 50 Years – the Continuing Relevance of Karl Popper*. London: Routledge.

Jones, W. T. (1975). *A History of Western Philosophy: The Twentieth Century to Wittgenstein and Sartre*. Vol. 4, *A History of Western Philosophy*. 2nd edn, revised. New York: Harcourt Brace Jovanovich.

Kahane, Howard (1990). *Logic and Philosophy: A Modern Introduction*. Belmont: Wadsworth.

Kant, Immanuel (1956). *Critique of Practical Reason*. Trans. Lewis White Beck. New York: The Bobbs-Merrill Company, Inc.

—— (1965). *Critique of Pure Reason*. Trans. Norman Kemp Smith. New York: St Martin's Press.

—— (1993). *Grounding for the Metaphysics of Morals*. Indianapolis: Hackett Publishing Company, Inc.

—— (2001). *Prolegomena to any Future Metaphysics*. Trans James W. Ellington. Indianapolis: Hackett Publishing Company, Inc.

Keuth, Herbert (1976), 'Verisimilitude or the Approach to the Whole Truth'. *Philosophy of Science* 43: 311–36.

—— (2005). *The Philosophy of Karl Popper*. Cambridge: Cambridge University Press.

Kirkham, Richard L. (1997). *Theories of Truth*. Cambridge, MA: The MIT Press.

Klausen, Søren Harnow (2004). *Reality Lost and Found: An Essay on the Realism–Antirealism Controversy*. University Press of Southern Denmark.

Kripke, Saul (1980). *Naming and Necessity*. Cambridge, MA: Harvard University Press.

—— (1982). *Wittgenstein on Rules and Private Language*. Cambridge, MA: Harvard University Press.

Kuhn, Thomas S. (1970). *The Structure of Scientific Revolutions* (2nd edn). Vol. 2, *International Encyclopedia of Unified Science*. Chicago: University of Chicago Press.

—— (1977). *The Essential Tension*. Chicago: University of Chicago Press.

—— (2000). *The Road Since Structure*. Chicago: University of Chicago Press.

Lakatos, Imre and Alan Musgrave (eds) (1968). *Philosophy of Science*. Amsterdam: North-Holland Publishing Company.

—— (eds) (1970). *Criticism and the Growth of Knowledge*. Cambridge: Cambridge University Press.

Laudan, L. (1980). 'A Refutation of Convergent Realism'. In *The Philosophy of Evolution*, ed. Uffe J. Jensen and Rom Hare, 232–68. New York: St Martin's Press.

Levinson, Paul (ed.) (1982). *In Pursuit of Truth*. Atlantic Highlands: Humanities Press Inc.

Lorenz, Konrad (1977). *Behind the Mirror*. Translated R. Taylor. New York: Harcourt Brace Jovanovich.

Mach, Ernst (1959). *The Analysis of Sensations*. Translated C. M. Williams. New York: Dover Publications.

Mackie, J. L. (1973). *Truth, Probability, and Paradox*. Oxford: Clarendon Press.

Magee, Bryan (1985). *Philosophy and the Real World*. La Salle: Open Court.

Malcolm, Norman (1986). *Nothing is Hidden: Wittgenstein's Criticism of His Early Thought*. Oxford: Basil Blackwell, Inc.

Martinich, A. P. (ed.) (1990). *The Philosophy of Language*. New York: Oxford University Press.

Meyer, Michel (ed.) (2001). *Questioning Derrida*. Aldershot: Ashgate.

Miller, David (1974). 'Popper's Qualitative Theory of Verisimilitude'. *British Journal for the Philosophy of Science* 25: 166–77.

—— (ed.) (1985). *Popper Selections*. Princeton: Princeton University Press.

—— (1994). *Critical Rationalism: A Restatement and Defense*. La Salle: Open Court.

—— (1995). 'Propensities and Probabilities'. In *Karl Popper: Philosophy and Problems*. British Society for the Philosophy of Science Series, ed. Anthony O'Hear, Oxford: Oxford University Press.

—— (2006). *Out of Error: Further Essays on Critical Rationalism*. Aldershot: Ashgate.

Misiek, Józef (ed.) (1995). *The Problem of Rationality in Science and Its Philosophy*. Dordrecht: Kluwer Academic Publishers.

Munz, Peter (1985). *The Knowledge of the Growth of our Knowledge*. London: Routledge and Kegan Paul.

—— (2004). *Beyond Wittgenstein's Poker: New Light on Popper and Wittgenstein*. Aldershot: Ashgate.

Musgrave, Allan (1971). 'Falsification and Its Critics'. *Proceedings of the Fourth International Congress for Logic, Methodology and the Philosophy of Science*.

—— (1999). *Essays on Realism and Rationalism*. Amsterdam: Rodopi.

Nagel, Ernest (1979). *Teleology Revisited*. New York: Columbia University Press.

Newton-Smith, W. H. (1981). *The Rationality of Science*. London: Routledge.

—— and Jiang Tianji (eds) (1992). *Popper in China*. With the assistance of E. James. London: Routledge.

Notturno, Mark (1984). 'The Popper/Kuhn Debate: Truth and Two Faces of Relativism'. *Psychological Medicine*.

—— (1985). *Objectivity, Rationality, and the Third Realm: Justification and the Grounds of Psychologism*. Dordrecht: Martinus Nijhoff.

—— (2000). *Science and the Open Society: The Future of Karl Popper's Philosophy*. Budapest: CEU Press.

—— (2003). *On Popper*. London: Thomson Wadsworth.

O'Hear, Anthony (1980). *Karl Popper*. London: Routledge and Kegan Paul.

—— (ed.) (1995). *Karl Popper: Philosophy and Problems*. British Society for the Philosophy of Science Series. Oxford: Oxford University Press.

Bibliography

Passmore, John (1967). 'Logical Positivism'. *The Encyclopedia of Philosophy*, ed. Paul Edwards. New York: Doubleday, Doran and Company.

Plato (1989). *Meno*. In *The Collected Dialogues of Plato*, ed. Edith Hamilton and Huntington Cairns. Princeton: Princeton University Press.

Platts, Mark (1997). *Ways of Meaning: An Introduction to a Philosophy of Language*. Cambridge, MA: The MIT Press.

Polanyi, Michael (1964). *Personal Knowledge: Towards a Post-Critical Philosophy*. New York: Harper & Row.

Popper, Karl (1945a/1971). *The Open Society and Its Enemies* Vol. I. Princeton: Princeton University Press.

—— (1945b/1971). *The Open Society and Its Enemies* Vol. II. Princeton: Princeton University Press.

—— (1957). *The Poverty of Historicism*. London: Routledge and Kegan Paul.

—— (1959a). *The Logic of Scientific Discovery*. New York: Basic Books, Inc.

—— (1959b). 'The Propensity Interpretation of Probability'. *British Journal for the Philosophy of Science* 10: 25–42.

—— (1963/1989). *Conjectures and Refutations* (5th rev. edn). London: Routledge.

—— (1979). *Objective Knowledge* (rev. edn). Oxford: Clarendon Press.

—— (1982a). *The Open Universe*. London: Routledge.

—— (1982b). *Quantum Theory and the Schism in Physics*. London: Unwin Hyman Ltd.

—— (1982c). *Unended Quest*. La Salle: Open Court.

—— (1983). *Realism and the Aim of Science*. London: Routledge.

—— (1987) 'Natural Selection and the Emergence of Mind'. In Radnitzky (1987), 139–55.

—— (1990). *A World of Propensities*. Bristol: Thoemmes.

—— (1996). *In Search of a Better World*. Trans. Laura J. Bennett. London: Routledge.

—— (1997). *The Myth of the Framework*. Ed. M. A. Notturno. London: Routledge.

—— (2001). *The World of Parmenides: Essays on the Presocratic Enlightenment*. Ed. Arne F. Petersen with assistance of Jørgen Mejer. London: Routledge.

—— (2005). *All Life is Problem Solving*. Trans. Patrick Camiller. London: Routledge.

—— and contributors (1974). *The Library of Living Philosophers*. Vol. 14, *The Philosophy of Karl Popper*, ed. Paul Arthur Schilpp. La Salle: Open Court.

—— and David W. Miller (1983). 'A Proof of the Impossibility of Inductive Probability'. *Nature* 302: 687–8.

—— and David W. Miller (1987). 'Why Probabilistic Support is not Inductive'. *Philosophical Transactions of the Royal Society*, 569–91.

—— and John C. Eccles (1977). *The Self and Its Brain*. Berlin: Springer International.

Preti, Consuelo (2003). *On Kripke*. Toronto: Thomson Wadsworth.

Purves, William K. and Gordon H. Orians (1984). *Life: The Science of Biology*. Boston: Willard Grant Press.

Putnam, Hilary (1983). 'Why Reason Can't Be Naturalized'. In *Realism and Reason: Philosophical Papers*, Vol. 3. Cambridge, MA: Cambridge University Press, 229–47.

—— (1989). 'Why Is a Philosopher?' In *The Institution of Philosophy*, ed. Avner Cohen and Marcelo Dascal, 73–86. La Salle: Open Court.

Quine, W. V. (1951) 'Two Dogmas of Empiricism'. In Gibson (2004).
—— (1970). *Philosophy of Logic*. New Jersey: Prentice-Hall.
—— (1982). *Methods of Logic*. Cambridge, MA: Harvard University Press.
Radnitzky, Gerard (ed.) (1987). *Evolutionary Epistemology, Rationality, and the Sociology of Knowledge*. La Salle: Open Court.
Raphael, Frederic (1999). *Popper*. New York: Routledge.
Rawls, John (1971). *A Theory of Justice*. London: Oxford University Press.
Reichenbach, Hans (1930). 'Kausalitat und Wahrscheinlichkeit'. *Erkenntnis 1*: 158–88.
Rorty, Richard (1980). *Philosophy and the Mirror of Nature*. Princeton: Princeton University Press.
—— (1982). *Consequences of Pragmatism*. Minneapolis: University of Minnesota Press.
—— (1989). *Contingency, Irony, Solidarity*. Cambridge: Cambridge University Press.
—— (1991). *Objectivity, Relativism, and Truth*. Cambridge: Cambridge University Press.
—— (1999). *Philosophy and Social Hope*. London: Penguin Books.
Ruse, Michael (1989). 'The View from Somewhere: A Critical Defense of Evolutionary Epistemology'. In *Issues in Evolutionary Epistemology*, ed. K. Hahlweg and C. A. Hooker, 185–228. Albany: SUNY Press.
Russell, Bertrand (1912). *The Problems of Philosophy*. London: Williams and Norgate.
—— (1936). 'The Limits of Empiricism'. In *Proceedings of the Aristotelian Society*, Vol. 36: 131–50. Oxford: Oxford University Press.
—— (1985). *The Philosophy of Logical Atomism*. Ed. David Pears. La Salle: Open Court.
Sceski, John H. (2000). *Popper on Objectivity*. Dissertation, Saint Louis University, 1999. Ann Arbor: UMI Dissertation Services.
Scruton, Roger (1988). *Kant*. Oxford: Oxford University Press.
Shearmur, Jeremy (1996). *The Political Thought of Karl Popper*. London: Routledge.
Simkin, Colin (1993). *Popper's Views on Natural and Social Science*. New York: E. J. Brill.
Sober, Elliot (1993). *The Philosophy of Biology*. Boulder: Westview Press.
—— (1984). *The Nature of Selection*. Cambridge, MA: MIT Press.
Stadler, Friedrich (ed.) (2004). *Induction and Deduction in the Sciences*. Dordrecht: Kluwer Academic Publishers.
Stokes, Geoffrey (1998). *Popper: Philosophy, Politics and Scientific Method*. Cambridge: Polity Press.
Suppes, Patrick (1974). 'Popper's Analysis of Probability in Quantum Mechanics'. In *The Philosophy of Karl Popper*, ed. Paul A. Schilpp, 760–74. La Salle: Open Court.
Tarski, Alfred (1944). 'The Semantic Conception of Truth and the Foundations of Semantics'. In *Philosophy and Phenomenological Research*, Vol. 4: 341–75.
—— (1956). *Logic, Semantics, Metamathematics*. Oxford: Clarendon Press.
Ter Hark, Michel (2004). *Popper, Otto Selz and the Rise of Evolutionary Epistemology*. Cambridge: Cambridge University Press.
Tichy, Pavel (1974). 'On Popper's Definitions of Verisimilitude'. In *British Journal for the Philosophy of Science*, 25: 155–60.
Tidman, Paul (1994). 'Conceivability as a Test for Possibility'. In *American Philosophical Quarterly* 31(4): 297–309.

Vetter, Hermann (1977). 'A New Concept of Vensimilitude'. In *Theory and Decision* 8: 369–375.
Von Mises, Richard (1957). *Probability, Statistics, and Truth*. New York: Macmillan.
Wagner, Michael F. (1991). *An Historical Introduction to Moral Philosophy*. Englewood Cliffs: Prentice-Hall.
Watkins J. W. N. (1974). 'The Unity of Popper's Thought'. In *The Philosophy of Karl Popper*, ed. Paul A. Schilpp, 371–412. La Salle: Open Court.
Wettersten, John R. (1992). *The Roots of Critical Rationalism*. Amsterdam: Rodopi.
Wittgenstein, Ludwig (1961). *Tractatus Logico-Philosophicus*, trans. D. F. Pears and B. F. McGuinness, with an introduction by Bertrand Russell. London: Routledge and Kegan Paul.
—— (1968). *Philosophical Investigations* (3rd edn). Trans. G. E. M. Anscombe. New York: Macmillan.

Index

(Personal names cited in the Index reference those used in the body of the text.)

Agassi, Joseph   15

Bartley, William W.
  on comprehensively critical rationalism (CCR)   10
  on criticism, justificational/non-justificational   8–9
  on crisis in rationalist integrity   11
  on deductive transmissibility   144–145
  on fact/value distinction   144–145
Buhler, Karl
  on functions of language   23

Campbell, Donald
  on hypothetical realism   113–114
Cantor, Georg   11
Carnap, Rudolf   50
Curie
  Eve   46
  Marie and Pierre   46

Darwin, Charles   11
Davidson, Donald   xii
  on conceptual relativism and conceptual schemes   24, 133–135
Deconstruction   3–4, 53
Deluty, Evelyn   55
Demarcation
  logical positivists   33–34
  Popper on (see Popper)
Derrida, Jacques   4–5
Descartes, René   11, 21

Einstein, Albert
  and difference from amoeba   17, 109
  disbelief in his own theories   65

Fallibility
  of theories   7–8, 43
  of organisms   17
  politics   127, 130
Feyerabend, Paul   133
Frege, Gottlob
  Morning star/Evening star distinction   37
  on the objective content of thought   114
Freudian analysis
  immune to criticism   7
Fries Trilemma   50

Galilei, Galileo   71
  and critical inquiry   75

Hegel, Georg Wilhelm   28
Homology   112
Howson, Collin
  on the propensity interpretation of probability   101–102
Hume, David
  critique of induction as background for Popper's anti-justificationism   5–6
  belief epistemology   63
  empirical theory of meaning   9
  induction problem of   5–6
    and learning from experience   44
    Popper's solution to the problem of   60–76
  inference from facts to values   144
  metaphysics   69, 72
Husserl, Edmund   21

Jarvie, Ian
    on methodological rules 45–47
justificationism 5–9, 11
    and scientific method 60–61

Kant, Immanuel
    contradiction-in-thought test 146
    ethics 139
    famous dictum 14
    on the a priori nature of knowledge 15
    on the problem of demarcation 35ff
    Popper's thought as the finishing touch to 18
    transcendental deduction 134
Keuth, Herbert
    on decisions in science 52
    on corroboration and the uniformity of nature 54
    on Popper's agreement with Hume's critique of induction 61
    on Popper's argument for indeterminism 74
Kirkham, Richard 78
Kripke, Saul xii
    on objectivity as intersubjectivity 55–60
    on Wittgenstein's paradox 56–59
Kuhn, Thomas 2
    on conceptual relativism 134
    on the distinction between revolutionary and normal science 138
    on untranslatability 133

Lamarckism 17, 108
Laplace, Pierre-Simon
    demon argument 73, 118

Marx, Karl
    on the autonomy of sociology 27, 129
    on historicism 28
Meyer, Michel 4
Mill, J. S.
    on psychologism 27, 129
Miller, David W. xi
    classifying statements as true vs. believing statements 64–66
    complacency in science 43
    comprehensively critical rationalism (CCR) 11, 142
    crisis in rationalist integrity 11
    establishing an inductive principle 62
    futility of attacking an argument's premises 141
    metaphysics as coat-tailing 71
    on verisimilitude 13, 63, 81–82
    perpetration of error vs. perpetuation of error 12
    Popper's separation of knowledge from learning 44
    rationality and anti-justificationism 6n.23, 144
    repeatability in science 54
    skepticism as methodological tool, 12n.41
    Tarski's theory of truth 79, 80
Mises von, Richard
    on the frequency theory of probability 95
Musgrave, Alan
    on Miller and believing statements vs. classifying statements 64–65

Neo-Darwinian evolution 11
    simulates Lamarckism 17, 108
    situational logic 105–109
    and teleology 110
Neurath, Otto 50
Newton, Isaac
    laws of addition of forces 93
    treatment of space/time 68
    predictability of theory of forces 102
Newton's problem 68–75

O' Hear, Anthony 102
ought implies can 145

Plato
    on objectivity 21
    essentialism 22
    theory of forms 120
Popper Karl, passim
    amoeba and Einstein 17
    asymmetry 47–49
    basic statements 49–50

Popper Karl, passim (*continued*)
  belief rejection of 11–12, 16
  comprehensive critical rationalism (CCR) 9–11
  conjectural nature of knowledge 6, 60ff
  controlled guesswork 6
  corroboration 76–78
  critical thinking vs. dogmatic thinking 6, 20
  cultural relativism 25, 130–135
  deductive theory of testing 37–40
  demarcation
    problem of 34
    general logical form 35–37
    solution to the problem of 38
  determinism/indeterminism 71–76
  empirical basis 49–54
  essentialism rejection of 6
  evolutionary epistemology 17, 103–109
  fallibilism 7, 11
  falsification and falsifiability 48–49
  Fries trilemma 50
  foundationism 8–11, 52
  historicism 26–27, 137–139
  language
    four functions 110–112
  law of diminishing returns 59, 142
  induction
    four problems of 62–69
    general solution to the problem of 60–62
    solution to the metaphysical problem of 69–76
  metaphysics
    evolutionary ontology 109–115
    realism 6, 71, 76
    'tottering old metaphysician' 22
    world three
      and evolutionary ontology 112–114
      interactions of 114
      objectivity of 115–121
  methodology general features of 6
  myth of the framework 24–25, 130–133

  objectivity, passim
    and intersubjective criticism 8, 14–18
    general account of 16
    problems of 18–29
      in political theory 127
      in science 40–41
      Kripke on 55–60
  Oedipus effect 64
  philosophical method 4–7
  plastic controls 110
  social engineering
    piecemeal vs. utopian 28, 128, 138
  society
    closed vs. open 26–28, 137
  totalitarianism 26, 137–138
  truth
    use of Tarski's theory 78–81
  verisimilitude
    Miller/Tichy critique 13, 81
    Popper's theory of 81–84
  world three (see metaphysics)
  zero method 136
Pascal, Blaise 4
psychologism
  Mill on (see Mill)
  Neurath on (see Neurath)
  Popper on (see Popper)

Quine Willard V.
  on analytic statements, 6
  against propositional content or meaning 121
  on untranslatability 133

repeatability problem of 54–55
Rawls, John
  reflective equilibrium 145
Rorty, Richard
  neo-pragmatism as influenced by Wittgenstein 2–3
  on Darwinism and justification 23
Russell, Bertrand
  on philosophy 1
  on the theory of reference 112
  on theory of types 78
Rutherford, Ernst 91

Shearmur, Jeremy
　on Popperian ethics   139–140
situational logic
　biological   108
　sociological role of   136–137
Strawson, P. F.   134

Tarski, Alfred
　on truth   78–80
　Popper's use of   78–81
Tichy, Pavel
　on verisimilitude   13, 81
truth holism   42

Vetter, Hermann   14

Urbach, Peter
　on the propensity theory of probability
　　101–102

Wittgenstein
　against propositional content   121
　and essentialism   1–2, 19
　Kripke on   56–60
　limits of philosophy   1
　meaning as use   2
　skeptical paradox   56